D1253740

GENETICS, EVOLUTION, AND DISEASE

GENETICS, EVOLUTION, AND DISEASE

A collection of papers from symposia from the Fifty-First Annual Meeting of the American Association of Physical Anthropologists
Eugene, Oregon, April 1-3, 1982

Editors

DENNIS H. O'ROURKE
Department of Anthropology
University of Utah, Salt Lake City

GLORIA M. PETERSEN
Harbor-UCLA Medical Center
Torrance, California

FRANCIS E. JOHNSTON
Department of Anthropology
University of Pennsylvania, Philadelphia

ALAN R. LISS, INC. • NEW YORK

These papers are being printed both as an issue of the American Journal of Physical Anthropology, Volume 62, Number 1, 1983, with Dr. Dennis H. O'Rourke, Dr. Gloria M. Petersen, and Dr. Francis E. Johnston as Guest Editors, and as a separate volume, Genetics, Evolution, and Disease, edited by Dr. Dennis H. O'Rourke, Dr. Gloria M. Petersen, and Dr. Francis E. Johnston.

Address all Inquiries to the Publisher
Alan R. Liss, Inc., 150 Fifth Avenue, New York, NY 10011

Printed in the United States of America.

Library of Congress Cataloging in Publication Data

Main entry under title:

Genetics, evolution, and disease.

"A collection of papers from symposia from the fifty-first annual meeting of the American Association of Physical Anthropologists, Eugene, Oregon, April 1-3, 1982."
Also issued as no. 1 of v. 62 of the American Journal of Physical Anthropology, 1983.
Includes bibliographical references and index.
1. Medical genetics--Congresses. 2. Human evolution --Congresses. 3. Human population genetics--Congresses. I. O'Rourke, Dennis H. II. Petersen, Gloria M. III. Johnston, Francis E. IV. American Association of Physical Anthropologists.
RB155.G388 1983 616'.042 83-14875
ISBN 0-8451-0227-3

Contents

Contributors

Peter H. Bennett [107] National Institute of Arthritis, Diabetes, and Digestive and Kidney Diseases, Phoenix, AZ 85014

R.M. Fineman [23] University of Utah Medical Center, Salt Lake City, UT 84132

C.J. Glueck [33] University of Cincinnati, College of Medicine, Cincinnati, OH 45267

Michael Grace [3] Alberta Cancer Board, Edmonton, Alberta, Canada

C.L. Gulbrandsen [33] University of Hawaii, Honolulu, HI 96822

T. Mary Holmes [3] University of Alberta, Edmonton, Alberta, T6G 2E9, Canada

L.B. Jorde [23] University of Utah Medical Center, Salt Lake City, UT 84132

A. Kagan [33] University of Hawaii, Honolulu, HI 96822

Sam Kemel [3] Manitoba Cancer Treatment and Research Foundation, Winnipeg, Manitoba, Canada

William C. Knowler [107] National Institute of Arthritis, Diabetes, and Digestive and Kidney Diseases, Phoenix, AZ 85014

P. Laskarzewski [33] University of Cincinnati, College of Medicine, Cincinnati, OH 45267

R.A. Martin [23] University of Utah Medical Center, Salt Lake City, UT 84132

M. McGue [33] Washington University, St. Louis, MO 63110

P. McGuffin [51] Institute of Psychiatry, Maudsley Hospital, London

Kenneth Morgan [3] University of Alberta, Edmonton, Alberta, T6G 2E9, Canada

N.E. Morton [33] University of Hawaii, Honolulu, HI 96822

William H. Mueller [11] The University of Texas, Houston, TX 77025

Dennis H. O'Rourke [1,51] University of Utah, Salt Lake City, UT 84112

Gloria M. Petersen [1,71] UCLA School of Medicine, Harbor-UCLA Medical Center, Torrance, CA 90509

Nicholas L. Petrakis [115] University of California, San Francisco, CA 94143

David J. Pettitt [107] National Institute of Arthritis, Diabetes, and Digestive and Kidney Diseases, Phoenix, AZ 85014

J.A. Pierce [19] Washington University School of Medicine, St. Louis, MO 63110

Ernesto Pollitt [11] The University of Texas, Houston, TX 77025

D.C. Rao [33] Washington University, St. Louis, MO 63110

T. Reich [51] Washington University School of Medicine, St. Louis, MO 63110

G.G. Rhoads [33] University of Hawaii, Honolulu, HI 96822

D.F. Roberts [67] University of Newcastle upon Tyne, Newcastle upon Tyne, NE2 4AA, United Kingdom

Diane Robson [3] Saskatchewan Cancer Foundation, Regina, Saskatchewan S4S 6X3, Canada

Jerome I. Rotter [71] UCLA School of Medicine, Harbor-UCLA Medical Center Medical Center, Torrance, CA 90509

J.M. Russell [33] Washington University, St. Louis, MO 63110

Warren Strober [119] National Cancer Institute, National Institute of Health, Bethesda, MD 20014

The number in brackets is the opening page number of the contributor's article.

B.K. Suarez [19] Washington University School of Medicine, St. Louis, MO 63110

Glenys Thomson [81] University of California, Berkeley, CA 94720

R.H. Ward [91] University of British Columbia, Vancouver, British Columbia V6T 1W5

Robert C. Williams [107] Arizona State University Tempe, AZ 85281

W.R. Williams [33] The University of Texas System Cancer Center, Texas Medical Center, Houston, TX 77030

Shozo Yokoyama [61] Washington University, St. Louis, MO 63110

AMERICAN JOURNAL OF PHYSICAL ANTHROPOLOGY 62:1–2 (1983)

Biological Anthropology and Genetic Disease Research: Introduction

DENNIS H. O'ROURKE AND GLORIA M. PETERSEN
Department of Anthropology, University of Utah, Salt Lake City, Utah 84112 (D.H.O.), and Division of Medical Genetics, Harbor-UCLA Medical Center, Torrance, California 90509 (G.M.P.)

Biological anthropologists have long been interested in the evolution of diseases and their impact on the course of human evolution itself. Recently, biological anthropologists have utilized their unique evolutionary perspective in approaches to disease etiology. In focusing on genetic disease, they share concerns with human geneticists, and encourage the use of newer, more sophisticated genetic analytic techniques. They are also afforded opportunities to explore data bases, particularly biomedical research-generated clinical data, formerly outside the anthropological domain. As methodology and technology in biomedicine have developed, so too have the nonclinical ramifications.

In an effort to emphasize this growing research area within our discipline, we organized symposia for the Fifty-First Annual Meeting of the American Association of Physical Anthropologists, held April 1–3, 1982, in Eugene, Oregon. One of our specific aims was to stress the potential of the interface between biological anthropology and genetic disease research. Although planned independently, every effort was made to make the symposia contents complementary. The diversity of approach and application underscores the extent to which biological anthropology can both contribute to and benefit from recent developments in genetic disease research. One symposium, Genetic Analysis of Human Disease: Application to Clinical and Population Data (organized by D.H.O.), was structured to emphasize the growing use of genetic analytic techniques and broadening of data bases which are characteristic of this field of inquiry. The other symposium, Genetic Disease Research: The Evolutionary Perspective (organized by G.M.P.), was designed to bring attention to the contributions that anthropological and evolutionary approaches can make to understanding genetic diseases.

Two themes can be traced through the papers presented in this issue: the interplay of biology and culture in the distribution of genetic disease, and the evolutionary implications of genetic diseases. Both themes are of special interest to anthropologists, and they merit consideration in genetic disease research.

The contributions of Morgan et al. and Mueller and Pollit utilize a population-based approach for the study of disease distribution and etiology. The study by Morgan et al. stresses the importance of careful epidemiologic investigations as preparatory to genetic analysis, by examining the patterns of cancer incidence in a Canadian religious isolate based on ascertainment through a population registry. Mueller and Pollitt present the novel use of the analysis of familial correlations for physical growth parameters to aid in assessing malnutrition.

In a further illustration of methods employed to examine genetic and environmental contributions to complex phenotypes, Rao and colleagues utilize the method of path coefficients to assess risk factors in coronary heart disease, and as a method for comparing the results of three independent family studies. Alternatively, Jorde et al. demonstrate the utility of large pedigree analysis, based on the Utah Genealogical Data Base, to examine the distribution and mode of transmission of neural tube disorders in a large, relatively homogeneous population.

The studies by O'Rourke et al. and by Suarez and Pierce illustrate not only the diversity of genetic analytic techniques available, but how well data, seldom employed by biological anthropologists, can be utilized. Suarez and Pierce take a distinctly genetic approach to study the transmission properties of a single locus that is associated with several related but discrete disease entities. Conversely, O'Rourke et al. use a variety of genetic analytic techniques to attempt to elucidate genetic mechanisms in the

Received August 5, 1982; accepted March 16, 1983.

formation of a complex behavioral phenotype.

There are significant evolutionary implications of genetic disease: the impact of genetic disorders on human evolution, the evolution of the disease itself, and its interaction with human culture are important issues for physical anthropology. The theoretical contribution by Yokoyama places the study of disease distribution and etiology within an evolutionary framework through the mechanism of social selection. The papers by Petersen and Rotter on peptic ulcer disease, and Strober on gluten-sensitive enteropathy, postulate an interaction between human culture and infectious agents on the increase in frequency of genes predisposing to disease states. Knowler examines the relationship of shifts in Pima Indian cultural practices as they affected the development of diabetes mellitus. Using data comparing the migrant and nonmigrant populations on Tokelau, Ward discusses the gentic, environmental, and cultural effects on high blood pressure. Petrakis suggests that the wet/dry cerumen polymorphism found in different racial groups may be an indicator of predisposition to breast disease. Finally, Thomson discusses the highly polymorphic HLA system as an important focus for analyzing disease associations and evolutionary patterns, using ankylosing spondylitis as an example.

In his contribution, Derek Roberts presents an overview of the issues presented in a number of the studies in this volume. The synthesis provided in his study is of major importance for our field. The cross-fertilization of biological anthropology and genetic epidemiology can only be mutually beneficial. Biological anthropology contributes a unique perspective from its traditional concerns with human population structure and evolutionary dynamics; it can benefit tremendously from the advanced analytic techniques and broader data bases. Further, by changing our perception of phenotype, and taking a more encompassing view of variation, more realistic and profitable approaches to the study of gene–environment interaction and covariation may be achieved.

Finally, two rather more practical benefits of this disciplinary cross-fertilization should be mentioned. First, genetic disease research is multidisciplinary almost by necessity. This is especially true in the area of applied medical research. The nature of interdisciplinary collaborative research fosters the exchange of ideas, approaches, and techniques which are frequently synergistic, such that the result is superior to that which could have been achieved by any single investigator. Although collaborative research is not new to biological anthropology, a concerted approach in genetic disease research provides many more opportunities for investigation.

Second, in a period of decline in funding for research in the traditional domain of physical anthropology, exploring areas in the biomedical sciences to which biological anthropologists can make significant contributions is appropriate. The greatest advances in scientific understanding were made by those who took the threads of diverse avenues of investigation and wove them into a broader synthesis. We believe the time is ripe to demonstrate the potential for biological anthropologists to be in the vanguard of this development.

AMERICAN JOURNAL OF PHYSICAL ANTHROPOLOGY 62:3–10 (1983)

Patterns of Cancer in Geographic and Endogamous Subdivisions of the Hutterite Brethren of Canada

KENNETH MORGAN, T. MARY HOLMES, MICHAEL GRACE, SAM KEMEL, AND DIANE ROBSON
Department of Genetics, University of Alberta, Edmonton, Alberta, T6G 2E9 (K.M., T.M.H.), Alberta Cancer Board, Edmonton, Alberta (M.G.) Manitoba Cancer Treatment and Research Foundation, Winnipeg, Manitoba (S.K.), and Saskatchewan Cancer Foundation, Regina, Saskatchewan, S4S 6X3, Canada (D.R.)

KEY WORDS Hutterite Brethren, Religious isolate, Cancer incidence, Smoking-associated cancers

ABSTRACT The Hutterite Brethren comprise a religious isolate and live on communal agricultural farms (colonies) in North America. In 1976 there were approximately 15,000 Canadian Brethren living in 179 colonies of the three endogamous subdivisions, the Dariusleut, Lehrerleut, and Schmiedeleut. Dariusleut and Lehrerleut colonies are located in both Alberta and Saskatchewan, and the Schmiedeleut are in Manitoba. Brethren were identified on population-based cancer registries of the three Prairie Provinces and among death registrations in the vital statistics of Alberta and Saskatchewan. The method of ascertainment was by a search for the 15 contemporary surnames and verification by address. 89 male and 91 female Brethren were identified who had cancer during the period, 1956–1975. The numbers of observed cancers were less than expected from provincial incidence rates for males and females in each province. The largest deficits were for female Brethren in Manitoba and Saskatchewan. There is a marked deficiency of cancer of the uterine cervix among female Brethren. In males there is a significant deficit of lung cancer. The Hutterite way of life contributes to a low risk for cancers of smoking-associated sites. However, there is evidence that male Brethren in Alberta may be at relatively increased risk for stomach cancer and leukemias. The site distribution patterns of cancers among the three endogamous leut are similar.

The Hutterite Brethren of North America are Anabaptists who live on communal farms, called colonies (Hostetler, 1974). There are only 15 extant surnames. Thus using both surname and address it is possible to identify Hutterite Brethren on population-based registries of, for example, cancer patients and vital statistics death registrations. This is an efficient procedure considering that the Brethren constitute a small minority. In the Prairie Provinces of Canada—Alberta, Saskatchewan, and Manitoba—there were 15,000 Hutterite Brethren in 1976, which is less than 0.4% of the population.

The Hutterite population is hierarchically structured with three endogamous subdivisions called *leut*—the Dariusleut, Lehrerleut, and Schmiedeleut. The leut view themselves as separate groups and there are only a very small number of interleut marriages. Residence is patrilocal and new colonies are formed by a fission, lineal process. An attempt is made to balance the age-sex distribution between the families who remain and those who take up residence in the daughter colony. Thus within each leut, colony lineages are established (Morgan and Holmes, 1982).

Received June 10, 1982; accepted March 16, 1983.

Dr. Grace's present address is: Department of Medicine, University of Alberta, Edmonton, Alberta T6G 2G3, Canada.

Dr. Kemel's present address is: B.C. Systems Corporation, Victoria, British Columbia V8T 4W9, Canada.

The Brethren have been in continuous residence in Canada since 1918. In 1976, 179 Hutterite colonies were located in Canada in the Prairie Provinces and 63 colonies were located in the United States in Washington, Montana, North Dakota, South Dakota, and Minnesota (Fig. 1). The geographic and endogamous subdivisions overlap in that the western borders of Manitoba, North Dakota, and South Dakota delimit the Dariusleut and Lehrerleut colonies to the west and Schmiedeleut colonies to the east.

The Brethren constitute a genetic isolate since their ancestry can be traced to 443 individuals on the 1880 census of the Dakota Territory, plus three large families with the surname "Tschetter" who joined the colonies between 1883 and 1892 (Eaton and Mayer, 1953) and a few migrants who subsequently married into the Hutterite population. The founders of each leut form partially overlapping sets of ancestors (Martin, 1970). Although there is variability among colonies in wealth and in agricultural specialization, the environment within a colony is relatively homogeneous.

The Brethren are an interesting group for epidemiologic studies because of their distinctive lifestyle, hierarchical population structure, and agrarian, communal society. The incidence of cancer in Canadian Hutterite Brethren in the 20-year period 1956–1975 is the subject of this report. This is an extension of the study of incidence of cancer in the Alberta Brethren for the period 1953–1974 (Gaudette et al., 1978).

MATERIALS AND METHODS

Population-based cancer registries are maintained by the Alberta Cancer Board of Alberta, the Saskatchewan Cancer Foundation, and the Manitoba Cancer Treatment and Research Foundation. The sources of data for the registries include physicians' referrals to cancer clinics, reports of hospital inpatients and outpatients, reports of physician office visits, pathology reports, and death certificates. Records have been kept on cancer patients in Alberta since 1941 and have been the basis of a population-based registry since 1951. Comprehensive registration of cancer patients in Sas-

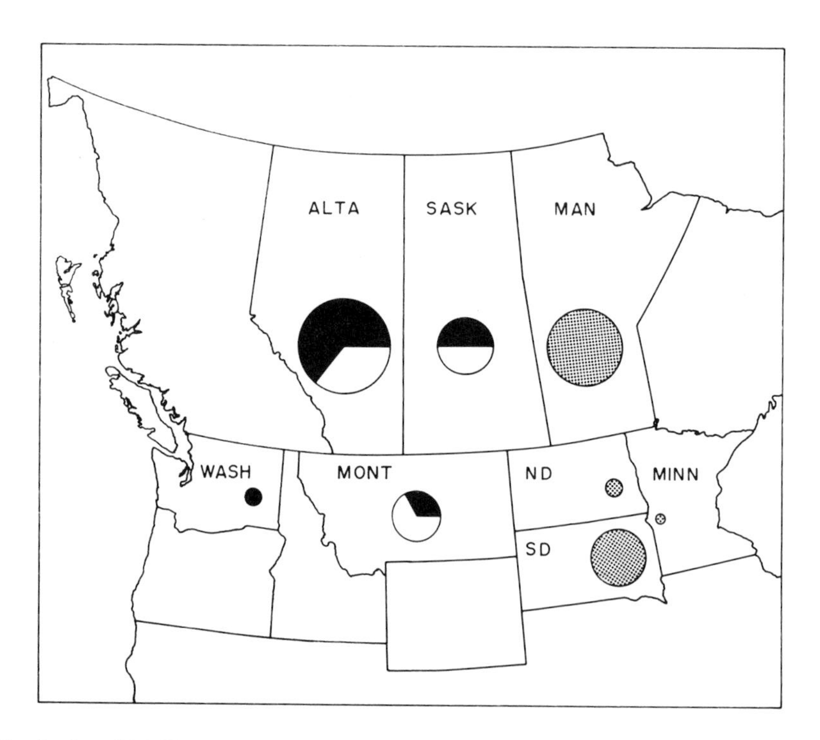

Fig. 1. Distribution of 242 Hutterite colonies in North America in 1976. The area of the pie chart is proportional to the number of Hutterite colonies. Solid sectors represent the number of Dariusleut colonies (83), open sectors represent the number of Lehrerleut colonies (63), and stippled circles represent the number of Schmiedeleut colonies (96).

katchewan began in 1944. A central registry for the enumeration of cancer cases in Manitoba was created in 1937 and was reorganized on a population basis in 1950 (Waterhouse et al., 1976).

The Hutterite Brethren were identified by searching the registries for patients with any of the 15 contemporary Hutterite surnames and then verifying Hutterite status, primarily by colony address. In addition, in both Alberta and Saskatchewan vital statistics death registrations were searched for Brethren dying of cancer who could not be identified on the cancer registries.

Four age-sex distributions of the Hutterite population were used in the calculation of person-years at risk: for 1953–1957, the average of the December 1950 distribution (Eaton and Mayer, 1953) and a composite distribution comprised of colony censuses obtained during the period 1953–1963 (Steinberg et al., 1967); for 1958–1963, the composite distribution; for 1964–1970, the average of the composite distribution and the June 1971 distribution of Hutterite Brethren in the Prairie Provinces (Statistics Canada, special tabulation from the 1971 Canadian census); and for 1971–1975, the June 1971 distribution.

Population sizes of Hutterite colonies are available for various years: 1951, 1956, 1961, 1966, 1971, 1976 (Statistics Canada, personal communication and special tabulations); 1964 (Peters, 1965); 1969 (Friedmann, 1970); 1972 (Alberta Select Committee of the Assembly, 1972); 1970, 1971, 1972, 1974, 1975 (Department of Municipal Affairs, Government of Saskatchewan, personal communication); 1968, 1970, 1975 (Ryan, 1977). Smoothed estimates of the population for June of each year, 1956–1975, and province were obtained from the linear regressions of the population sizes on year ($r^2 = 0.97$, 0.91, and 0.98 for Alberta, Saskatchewan, and Manitoba, respectively.) Hutterite person-years at risk for each province were calculated by multiplying the appropriate age-sex distribution and the smoothed population estimates and cumulating over the 20 years.

Tabulations of cancer cases by site, age, and sex are available for each of the population-based registries. Annual age-by-sex distributions of the population of the provinces are published (Statistics Canada, 1973, 1979). Cancer incidence rates for each province, averaged over the study period, were computed from the tabulation of cancer cases and the provincial population. The expected numbers of Hutterite cancer cases were computed by multiplying the provincial incidence rates and the Hutterite person-years at risk for the respective provinces. Alberta and Saskatchewan provincial incidence rates do not include cancer cases ascertained only through vital statistics death registrations. The cancers are identified by site according to Eighth Revision International Classification of Diseases, Adapted for Use in the United States (ICDA) (U.S. Department of Health, Education and Welfare, 1968).

The significance of observed compared to expected numbers of cancers was assessed by inspection of the Freeman-Tukey deviates (Bishop et al., 1975; Sokal and Rohlf, 1981). The deviates are calculated according to the formula $z = \sqrt{x} + \sqrt{x+1} - \sqrt{4m+1}$, where x is the observed number and m is the expected number of cancer cases. This formula results from considering the variance stabilizing transformation $y = \sqrt{x} + \sqrt{x+1}$. Let x be a Poisson variable with mean m, then y is approximately normally distributed, with approximate mean $\sqrt{4m+1}$ and variance 1.

RESULTS AND DISCUSSION

Comparison by province

During 1956–1975, there were 89 male Brethren with a total of 93 cancers and 91 female Brethren with 93 cancers; 4 males and 2 females each had two primary cancers. The number of Hutterite cancer cases ascertained from the cancer registry in each province is presented in Table 1. (Cases ascertained only from Alberta vital statistics death registrations are not included in Table 1. There were no Brethren who died of cancer in Saskatchewan who were not on the cancer registry.) Because "other cancer of the skin" (ICDA 173), which is not a malignant melanoma, appeared not to be consistently ascertained by the three registries, comparison of all sites excluding ICDA 173 is considered more meaningful. The ratios of observed to expected cancers for males are lower for Saskatchewan and Manitoba than for Alberta. For females in all three provinces, the observed number of cancers was much less than expected. It should be noted that the average incidence rates differ among the three provinces during the time period of this study.

A specific cancer site was chosen for further analysis if the number of expected or observed cases was at least five for either sex. Freeman-Tukey deviates were calculated to assess the significance of observed compared to expected numbers of cancer cases. An absolute value exceeding 1.96 for any given comparison was

TABLE 1. Observed and expected number of primary cancer cases among Hutterite Brethren, 1956–1975

ICDA	Cancer Site or Type	Alberta		Saskatchewan		Manitoba	
		Obs	Exp	Obs	Exp	Obs	Exp
				Males			
	All sites	47	57.7	13	18.0	28	48.0
	All sites excluding other skin	43	45.2	9	13.8	25	40.6
140	Lip	2	3.1	0	1.1	1	1.7
151	Stomach	8	3.0	1	0.9	4	2.9
153	Large intestine except rectum	1	3.4	2	0.8	0	3.1
154	Rectum and rectosigmoid junction	5	2.3	0	0.8	3	1.8
162	Trachea, bronchus, and lung	0	5.7	0	1.6	2	6.0
173	Other skin	4	12.4	4	4.1	3	7.3
185	Prostate	7	4.8	1	1.7	1	3.8
188	Bladder	1	2.7	0	0.7	0	1.9
204–207	Leukemias	7	2.4	3	0.8	3	2.3
				Females			
	All sites	51	72.9	9	19.7	24	60.6
	All sites excluding other skin	43	63.1	8	16.8	24	55.6
151	Stomach	3	1.3	1	0.4	3	1.3
153	Large intestine except rectum	1	3.9	0	1.0	3	3.2
173	Other skin	8	9.8	1	2.9	0	5.0
174	Breast	18	15.4	3	4.1	8	11.2
180 and 234.0	Cervix uteri	1	17.3	0	2.9	2	15.7
181 and 182	Chorionepithelioma and other uterus	3	3.8	1	1.2	1	2.8
183	Ovary, fallopian tube, and broad ligament	1	2.8	0	0.8	1	1.5

considered significant at the 5% level (two-tailed). The Freeman-Tukey deviates (Table 2) indicate that the excesses of stomach cancer (ICDA 151) and of leukemias (ICDA 204–207) are statistically significant only for Alberta males. There is a marked deficiency of lung cancer (ICDA 162) in men, with the Alberta Brethren showing the greatest deficiency. For women there is a marked deficiency of invasive cancer and carcinoma in situ of the uterine cervix (ICDA 180 and 234.0). This is consistent with an expected lower risk attributable to the strict mores of Hutterite sexual behavior (Kessler, 1977; Hulka, 1982). Disregarding other skin cancer (ICDA 173), there is no other significant excess or deficiency of cancer of a specific site in females.

Winkelstein (1977) proposed the hypothesis that, in addition to those cancers known to be associated with risk due to cigarette smoking, smoking also increases the risk of squamous cell cancer of the uterine cervix. Wigle et al. (1980) provided additional evidence for a dosage effect of smoking on the risk for uterine cervical cancer. More recently, Clarke et al. (1982) reviewed the literature on studies of the association between cigarette smoking and cervical dysplasia, carcinoma in situ, and invasive carcinoma of the uterine cervix. They report a case-control study confirming a two-fold risk for current smokers for cervical cancer after adjusting for age, socioeconomic status, and certain aspects of sexual behavior. The association between cigarette smoking and cancer of the uterine cervix was limited to squamous cell carcinoma. Clarke and colleagues suggest that a substance absorbed from cigarette smoke is secreted by the cervical epithelium where it may act as a promoter or cocarcinogen.

Since smoking is discouraged among the Brethren, sites associated with smoking risk were grouped in an attempt to explain the deficit of cancers. The cancers known or suspected to be associated with smoking are (Wigle et al., 1980): lip, tongue, mouth and pharynx, esophagus, pancreas, respiratory, cervix uteri, bladder, other urinary, and non-Hodgkin's lymphomas (ICDA 140, 141, 143–149, 150, 157, 160–163, 180 and 234.0, 188, 189, and 200 and 202, respectively). The numbers of observed and expected Hutterite cancers for smoking-associated sites and all other cancers (except other skin) are presented in Table 3 for each province. It is clear that the Brethren of both sexes are at decreased risk for cancers associated with smoking. But they do not appear to be consistently different from the general

TABLE 2. Freeman-Tukey deviates of observed compared to expected numbers of cancer cases for selected sites

ICDA	Cancer Site or Type	Alberta	Saskatchewan	Manitoba
			Males	
140	Lip	−0.5	−1.3	−0.4
151	Stomach	2.2	0.3	0.7
153	Large intestine except rectum	−1.4	1.1	−2.7
154	Rectum and rectosigmoid junction	1.5	−1.0	0.9
162	Trachea, bronchus, and lung	−3.9	−1.7	−1.9
173	Other skin	−2.9	0.1	−1.8
185	Prostate	1.0	−0.4	−1.6
188	Bladder	−1.0	−0.9	−1.9
204–207	Leukemias	2.2	1.7	0.5
			Females	
151	Stomach	1.2	0.8	1.2
153	Large intestine except rectum	−1.7	−1.2	0.0
173	Other skin	−0.5	−1.1	−3.6
174	Breast	0.7	−0.4	−0.9
180 and 234.0	Cervix uteri	−6.0	−2.5	−4.8
181 and 182	Chorionepithelioma and other uterus	−0.3	0.0	−1.1
183	Ovary, fallopian tube, and broad ligament	−1.1	−1.0	−0.2

TABLE 3. Observed and expected number of smoking-associated cancer cases among Hutterite Brethren, 1956–1975[1]

	Alberta		Saskatchewan		Manitoba	
Cancer Site	Obs	Exp	Obs	Exp	Obs	Exp
			Males			
Smoking-associated sites	4	18.5	1	5.6	6	16.0
All sites, excluding smoking-associated sites and other skin	39	26.8	8	8.3	19	24.7
			Females			
Smoking-associated sites	5	22.5	0	4.7	2	22.0
All sites, excluding smoking-associated sites and other skin	38	40.6	8	12.1	22	33.6

[1]Expected incidence of smoking-associated cancers was based on the average incidence for 1956–1975 in Manitoba; and for 1953–1975 for Alberta and Saskatchewan, adjusted for the person-years at risk for the period 1956–1975.

population with respect to risk of cancer at the other sites combined.

Comparison by leut

The observed number of total cancer cases for each leut (Table 4a) includes an additional five cases in males and nine in females that were ascertained from vital statistics death certificates in Alberta. These individuals were not included in the previous analysis of Alberta cancer registry cases because individuals ascertained only through vital statistics death registrations were not included in the average incidence rates for Alberta. There is no significant heterogeneity in the distribution of observed cases by leut and age-at-diagnosis category (for males, $X^2 = 7.47$; for females, $X^2 = 3.89$; df = 6). The four age-at-diagnosis categories are 0–24, 25–54, 55-74, and 75 years and older.

The heterogeneity of the site-distribution pattern of cancer among the leut was tested by contingency table analysis for 25 sites for males and 26 sites for females. For males, the profile of cancer by site does not differ significantly among the leut ($X^2 = 41.8$, df = 48). The profile for females appeared not to be significantly heterogeneous among the leut ($X^2 = 55.1$, df = 50); however, there are relative deficits of other skin cancer (ICDA 173) in Schmiedeleut females ($z = -2.21$) and of large intestine cancer (ICDA 153) in Dariusleut females ($z = -2.06$). Both effects are significant at the 5% level by the likelihood ratio test statistic (Bishop et al., 1975) ($G^2 = 5.70$ for ICDA 173; $G^2 = 5.36$ for ICDA 153, after adjustment for ICDA 173.)

TABLE 4a. Distribution of cancers among the leut, 1956–1975

Cancer Site or Type	Dariusleut	Lehrerleut	Schmiedeleut
		Males	
Smoking-associated sites	2	4	5
Other skin (ICDA 173)	4	4	3
Ill-defined sites (ICDA 195–199)	1	1	4
Leukemias (ICDA 204–207)	3	7	3
All other sites	21	18	13
All sites	31	34	28
		Females	
Smoking-associated sites	5	2	2
Other skin (ICDA 173)	4	5	0
Ill-defined sites (ICDA 195–199)	0	2	1
Leukemias (ICDA 204–207)	3	2	1
All other sites	27	19	20
All sites	39	30	24

TABLE 4b. Freeman-Tukey deviates

Cancer Site or Type	Dariusleut	Lehrerleut	Schmiedeleut
		Males	
Smoking-associated sites	−0.81	0.10	0.91
Other skin (ICDA 173)	0.28	0.10	−0.04
Ill-defined sites (ICDA 195–199)	−0.59	−0.71	1.37
Leukemias (ICDA 204–207)	−0.55	1.00	−0.35
All other sites	0.89	−0.18	−0.63
		Females	
Smoking-associated sites	0.67	−0.41	−0.06
Other skin (ICDA 173)	0.22	1.13	−2.21
Ill-defined sites (ICDA 195–199)	−1.46	0.94	0.39
Leukemias (ICDA 204–207)	0.41	0.19	−0.27
All other sites	−0.08	−0.45	0.74

Because of the small number of cancer cases for a number of sites and taking into account the results of comparisons by province, the cancer sites were grouped into a few larger categories and the data reanalyzed for heterogeneity among the leut (Table 4a,b). As with the previous analysis, the profile of cancer is not significantly different overall (for males, $X^2 = 8.46$; for females, $X^2 = 8.07$; df = 8). In particular, the leut profiles do not differ significantly with respect to smoking-associated sites nor leukemias. Note that cancer of the large intestine is included in the aggregate category, "all other sites." As previously noted, the absence of other skin cancer in Schmiedeleut females is statistically significant. Ascertainment of cases of other skin cancer and cancers of ill-defined sites are most likely to be affected by variability in registration procedures between registries and over time.

CONCLUSION

In our previous study of Alberta Brethren we found a significant overall deficit of cancers in females but no significant deficit in males (Gaudette et al., 1978). In Saskatchewan and Manitoba there are even fewer observed cancers than expected, especially for females. The deficit of observed cancers is largely accounted for by a deficiency of cancers of smoking-associated sites. The Hutterite woman is at low risk for cancer and at greatly reduced risk for uterine cervical cancer because of strict mores relating to sexual behavior in addition to absence of cigarette smoking.

Martin and colleagues (1980a) have reported on the cancer mortality of the Hutterite Brethren, primarily the Lehrerleut and Schmiedeleut, including Schmiedeleut Brethren from South Dakota. Cancer data on the Alberta and

Manitoba Brethren included in the mortality analysis are from the study reported here. Overall, the Hutterite Brethren had significantly fewer deaths from cancer than expected, utilizing U.S. mortality rates.

The pattern of cancer incidence in the Alberta Brethren shares a number of features with the pattern of cancer mortality in farmers (Gaudette et al., 1978). For example, in a study of death certificates in Iowa, male farmers had significantly lower mortality for smoking-related cancers (Burmeister, 1981). However, mortality rates were higher for the following six types of cancers: lip, stomach, leukemia, lymphomas, multiple myeloma, and prostate.

In addition to the rural agrarian environment there may be additional lifestyle factors as well as familial factors that influence risk for neoplastic disease in the Hutterite Brethren. These factors are most likely not homogeneously distributed in space or in time. For example, the Brethren migrated from the Ukraine, an area of high risk for stomach cancer (Haenszel, 1975). Migrants show rates of stomach cancer similar to their country of origin although there is some reduction in rates. In addition, the rate of stomach cancer has been declining. In Canada the mortality rate of stomach cancer has steadily declined from 1931 to 1972 and had the largest decrease in age-standardized rates of any particular site (Health and Welfare Canada, 1977). In this context it is significant that a familial cluster of stomach, digestive, and other cancers in the Dariusleut has been ascertained (Fowlow, 1973; Martin et al., 1980b).

The structure of the Hutterite population (Morgan and Holmes, 1982) poses methodological challenges for genetic epidemiologic studies among the Brethren. Because of patrilocal residence and the tendency toward lineal fissioning of colonies, brothers may spend their entire lives in the same colony or sequence of colonies. Although there is some opportunity for occupational specialization on the colony, Brethren residing within a colony will experience very similar environments. The result is a great degree of overlap in life experience for siblings of the same sex. Furthermore, there is the environment common to all members of the same colony.

Although the Brethren tend to avoid close consanguinity—and first cousin marriages are considered to be too close—multiple marriages occur between sibships (Bleibtreu, 1964; Mange, 1964). Therefore even in the situation of colony-exogenous marriages, married sisters could reside in the same colony because their husbands are related as brothers. Finally, because of overlapping sets of common ancestors in the founding of the three endogamous leut, the Hutterite population, at a more inclusive level, constitutes a very large genealogy for the genetic analysis of multiple, familial clusters of the same disease in the different leut.

ACKNOWLEDGMENTS

A version of this paper was presented at the 51st Annual Meeting of the American Association of Physical Anthropologists in Eugene, Oregon, April 1982. We gratefully acknowledge the assistance of J. Birdsell, V. Cloarec, L. Gaudette, H. Hersom, L. Laing, T. Snodgrass, and D. Wigle. This research was supported in part by the General Research Fund, University of Alberta.

LITERATURE CITED

Alberta Select Committee of the Assembly (Communal Property) (1972) Report on Communal Property 1972. Edmonton, Alberta: Queen's Printer.

Bishop, YMM, Fienberg, SE, and Holland, PW (1975) Discrete Multivariate Analysis: Theory and Practice. Cambridge, MA-MIT Press.

Bleibtreu, HK (1964) Marriage and Residence Patterns in a Genetic Isolate. Ph.D. Thesis. Cambridge: Harvard University.

Burmeister, LF (1981) Cancer mortality in Iowa farmers, 1971–78. J Natl Cancer Inst *66:*461–464.

Clarke, EA, Morgan, RW, and Newman, AM (1982) Smoking as a risk factor in cancer of the cervix: Additional evidence from a case-control study. Am J Epidemiol *115:*59–66.

Eaton, JW, and Mayer, AJ (1953) The social biology of very high fertility among the Hutterites. The demography of a unique population. Hum Biol *25:*206–264.

Fowlow, SB (1973) A Medical-Genetic Survey of a Human Isolate. M.Sc. Thesis. Calgary, Alberta: University of Calgary.

Friedmann, R (1970) A Hutterite census for 1969: Hutterite growth in one century, 1874–1969. Mennonite Q Rev *44:*100–105.

Gaudette, LA, Holmes, TM, Laing, LM, Morgan, K, and Grace, MGA (1978) Cancer incidence in a religious isolate of Alberta, Canada, 1953–74. J Natl Cancer Inst *60:*1233–1238.

Haenszel, W (1975) Migrant studies, In JF Fraumeni, Jr (ed): Persons at High Risk of Cancer: An Approach to Cancer Etiology and Control. New York: Academic Press, pp. 361–371.

Health and Welfare Canada (1977) Cancer Patterns in Canada, 1931–1974, Ottawa: Laboratory Centre for Disease Control.

Hostetler, JA (1974) Hutterite Society. Baltimore: Johns Hopkins University Press.

Hulka, BS (1982) Risk factors for cervical cancer. J Chron Dis *35:*3–11.

Kessler, II (1977) Venereal factors in human cervical cancer. Evidence from marital clusters. Cancer *39:*1912–1919.

Mange, AP (1964) Growth and inbreeding of a human isolate. Hum Biol *36:*104–133.

Martin, AO (1970) The founder effect in a human isolate:

Evolutionary implications. Am J Phys Anthropol *32:*351–367.

Martin, AO, Dunn, JK, Simpson, JL, Olsen, CL, Kemel, S, Grace, M, Elias, S, Sarto, GE, Smalley, B, and Steinberg, AG (1980a) Cancer mortality in a human isolate. J Natl Cancer Inst *65:*1109–1113.

Martin, AO, Dunn, JK, Simpson, JL, Elias, S, Sarto, GE, Smalley, B, Olsen, CL, Kemel, S, Grace, M, and Steinberg, AG (1980b) Genetics of neoplasia in a human isolate. In HV Gelboin, B MacMahon, T Matsushima, T Sugimura, S Takayama, and H Takebe (eds): Genetics and Environmental Factors in Experimental and Human Cancer. Tokyo: Japan Scientific Societies Press, pp. 291–302.

Morgan, K, and Holmes, TM (1982) Population structure of a religious isolate: The Dariusleut Hutterites of Alberta. In MH Crawford and JH Mielke (eds): Current Developments in Anthropological Genetics, Vol 2: Ecology and Population Structure. New York: Plenum Press, pp. 429–448.

Peters, V (1965) All Things Common: The Hutterian Way of Life. Minneapolis: University of Minnesota Press.

Ryan, J (1977) The Agricultural Economy of Manitoba Hutterite Colonies. Toronto: McClelland and Stewart.

Sokal, RR, and Rohlf, FJ (1981) Biometry, 2nd ed. San Francisco: WH Freeman and Company.

Statistics Canada (1973) Population 1921–1971: Revised Annual Estimates of Population, by Sex and Age Group, Canada and the Provinces. Catalogue 91-512. Ottawa: Information Canada.

Statistics Canada (1979) Population: Revised Annual Estimates of Population, by Sex and Age for Canada and the Provinces, 1971–1976. Catalogue 91-518. Ottawa: Information Canada.

Steinberg, AG, Bleibtreu, HK, Kurczynski, TW, Martin, AO, and Kurczynski, EM (1967) Genetic studies on an inbred human isolate. In JF Crow and JV Neel (eds): Proceedings of the Third International Congress of Human Genetics. Baltimore: John Hopkins Press.

U.S. Department of Health, Education and Welfare (1968) Eighth Revision International Classification of Diseases, Adapted for Use in the United States. Washington, D.C.:U.S. Government Printing Office.

Waterhouse, JAH, Muir, CS, Correa, P, and Powell, J (eds) (1976) Cancer Incidence in Five Continents, Vol III. Lyon, France: International Agency for Research on Cancer.

Wigle, DT, Mao, Y, and Grace, M (1980) Letters to the editor: Re: Smoking and cancer of the uterine cervix: Hypothesis. Am J Epidemiol *111:*125–127.

Winkelstein, W, Jr (1977) Smoking and cancer of the uterine cervix: Hypothesis. Am J Epidemiol *106:*257–259.

AMERICAN JOURNAL OF PHYSICAL ANTHROPOLOGY 62:11–17 (1983)

The Bacon Chow Study: Genetic Analysis of Physical Growth in Assessment of Energy–Protein Malnutrition

WILLIAM H. MUELLER AND ERNESTO POLLITT
The University of Texas, School of Public Health, Houston, Texas 77025

KEY WORDS Genetics, Growth, Nutrition

ABSTRACT Estimates of the prevalence of energy–protein malnutrition almost universally employ physical growth measurements. In this study we focus on this disease and the role of body size of relatives as mediators of responses in individuals to one type of nutrition intervention: supplementation of pregnant and lactating women. In this study, initiated by Dr. Bacon Chow and others in 1967, during gestation of a first infant a mother was untreated, while during the lactation of the first infant and the gestation and lactation of a second infant she was treated with either a calorie supplement or a placebo. Supplement–placebo group differences were sought in sibling and mother–child correlations in growth from birth to 30 months, in order to assess the role of heredity as a mediator of supplement effects.

There were 108 pairs of siblings whose mothers had received a high-calorie–high-protein supplement as described above and 105 pairs of siblings whose mothers had received a placebo. Among the latter, sibling correlations for most measurements are statistically significant at birth, and of the same magnitude seen in previous studies (~0.5), while among supplemented siblings, birth correlations are unusually low and often insignificant. The sibling correlations in Rohrer's index (wt/L^3) differed the most between groups ($p < 0.01$). Group differences in the sibling correlation tended to disappear over the first 2.5 years of life. Correlations between mothers and their second children in subscapular skinfold tended to be higher in the supplemented than in the placebo group, birth to 30 months. In *both* supplement groups mother–second child correlations for body weight were higher than mother–first child correlations, suggesting the occurrence of secular changes in the environment unconnected with the treatment. The results suggest that: (1) genetic analysis of components of anthropometric variation may be a more sensitive method than the more conventional comparison of group means in detecting supplement effects; and (2) infant relative weight (Rohrer index), particularly the addition of subcutaneous fat, may be more affected by maternal supplementation than growth in weight or length alone.

Genetic analyses of human disease usually focus on chronic disease in which biological factors may have a presumed important role. This study, however, focuses on a disease of ecological origin. Although genes do not cause energy–protein malnutrition, estimates of the prevalence of this problem as well as the usefullness of interventions to solve the problem, almost universally employ anthropometric measurements of physical growth. These measurements in turn have a known genetic basis, in part, even in nutritionally stressed populations (Russell, 1976; Martorell et al., 1977; Mueller, 1977; Mueller and Titcomb, 1977). Also the size and body composition of the mother and the intrauterine environment are known to affect birth size of infants (Robson, 1978). Few studies are available to address the role of body size of relatives as mediators of responses in individuals to nutrition intervention. One data set that is available, which can

Received June 10, 1982, accepted March 16, 1983

in part address the problem, is the study of nutritional supplementation of pregnant and lactating mothers in a rural population in Taiwan, initiated in 1967 by Dr. Bacon Chow and his co-workers. Due to the early death of Dr. Chow in September 1973 the analysis of the data from this study was delayed. In 1978 the second author (E.P.) obtained the data tape from the staff of the Nutrition Division of the United States Agency for International Development. Analyses of this tape were initiated with the inclusion of researchers who had collaborated with Dr. Bacon Chow at different stages of the original study (Blackwell et al., 1973; Chow, 1974; Herriott et al., 1977).

This study was unique in that each mother provided two participant infants. During gestation of the first infant the mother was untreated, while in the second gestation, she was treated with either a high-calorie supplement or a placebo. From these family data one can address several questions about the effectiveness of this kind of nutrition intervention: How may supplementation effects on growth be dependent on size of a previous sibling (reflecting genetic potential and/or conditions in utero)? How does maternal size mediate the effects of supplementation? In this paper we seek supplement-placebo group differences in sibling-sibling and mother-child anthropometric correlations, in an effort to address the above questions.

MATERIALS AND METHODS

Study design

This is a double-blind trial of nutrition supplementation of pregnant and lactating women. Two hundred ninety-four women were randomly assigned to one of two treatment groups (A and B). All women received a 12.5-ounce can of liquid supplement twice a day. The daily supplement for group A provided 800 kcal and 40 g of protein. Group B (the controls) received a liquid which resembled the A supplement in taste, texture, and weight. During the first 4 years of the study each can of placebo contained a total of 6 kcal/day. However, in June 1971 the artificial sweeteners in the placebo were replaced with sucrose, increasing the caloric content to about 40 kcal/per can (see Table 1). Moreover, vitamins and minerals were also added during the last year of the study. A tablet containing vitamins and minerals was delivered to the women in both groups each day. Treatment began 3 weeks after the delivery of a first infant. This infant forms the first (older) member of the sibling pairs dealt with in this study. Supplementation of the mothers continued during the lactation of this infant and throughout the gestation and lactation of a second infant—the younger member of the sibling pair. The supplement was not given to the children directly.

TABLE 1. Nutrient content of one 12.5-ounce can of the supplement for groups A and B

	Group A	Group B[1]	
		Before 6/71	After 6/71
Protein	20 g	—[2]	—[2]
Fat	13.3 g	—[2]	—[2]
Carbohydrates	50 g	—[2]	10 g
Calories	400 kcal	3 kcal	43 kcal

[1]76% of group B second study infants were born before 6/71.
[2]Trace amount.

The 294 women for this study were recruited in 14 villages in Sui-Lin township, a farming area about 180 miles from Taipei, Taiwan. They were judged to be in the lowest ranks of the socioeconomic order and to be nutritionally at risk, because a preliminary dietary survey showed that the average caloric intake was as low as 1,200 kcal with protein intake of less than 40 g per day. The staple diet consisted primarily of sweet potatoes and rice with little animal protein. Selection criteria also included having at least one normal male child, being in the second or third trimester of pregnancy, and planning to have at least one more child. The original age selection criterion called for inclusion of women between 20 and 28 years of age (Blackwell et al., 1973). However, the subjects recruited for the study ranged from 19 to 30 years of age. The population is one nutritionally at risk, because at birth Sui Lin infants are within the normal percentiles for weight and length of U.S. infants, but by 1 year of age, Sui Lin medians are close to the fifth percentile of the reference (Wohlleb et al., in press).

Of the 294 women recruited, 225 gave birth to two infants within the 6.5 years spanned by the study. The 69 who failed to meet this minimum requirement were evenly divided between groups A and B. The present analysis was restricted to those cases with anthropometric measurements on both infants within 48 hr of birth. Accordingly, six cases in each of the two groups were eliminated. The sample thus consists of 108 A group and 105 B group mother-child and sibling pairs.

Data collection and analysis

The data were collected in the field by teams of resident nurses who visited the families periodically throughout the study, and delivered the supplement twice daily. Compliance in consuming the supplement was monitored by the nurses who observed the subjects drinking the supplement or placebo and who then measured the amount remaining in the can after each delivery. Neither the subject nor the field investigators knew the nutrient contents of the respective cans.

Body measurements were taken at intervals from birth through 30 months of age. Some children were measured beyond 30 months, but sample sizes decrease markedly after this age. Children were measured within 48 hr of their birthdate. The anthropometrics were checked for any obviously erroneous values at the end of each day. Sample sizes decrease only slightly over the period from birth to 18 months. Loss in sample size with age was primarily due to the exclusion of individuals for the reasons cited earlier, or because of missing or clearly incorrect anthropometric measurements at one time period. If one member of a related pair had no measurements taken at a given age, the pair was excluded for that age comparison, but was included at all other ages for which measurements were complete.

Weight; length; head, chest, and abdomen circumferences; and triceps and subscapular skinfold measurements were obtained on each child. For the sibling correlations we have excluded chest and abdomen circumferences and triceps skinfold as we found these measurements highly correlated with body weight and subscapular skinfold, respectively. An index of relative weight, the Rohrer Index (wt/L^3), was also computed following the suggestion by Brandt (1979) that this index best reflects intrauterine nutrition. Mother's measurements include height, weight, and triceps and subscapular skinfolds. For the latter three measurements we had several choices as these were taken at repeated intervals over the lactation and pregnancy periods. For simplicity we chose those measurements one month after the birth of the child in question (either first or second), in the calculation of the mother–child correlations. This would include some information on the amount gained during pregnancy and would thus have potential nutritional significance.

Hypotheses. We would expect in this study, if supplementation of mothers has had an effect on child growth, to see a reduction in sibling covariation at birth in the A but not in the B group. If such a decrease occurs then genotype–environment interaction or environmental covariation is likely in physical (resemblance) of siblings, via the treatment effect imposed. A parallel question would focus on the extent to which any changes in sibling–sibling resemblance persist postnatally.

The mother–child relationship is more complex. The response might be different for the different anthropometric measurements of the mother. We might, for example, expect a reduction in the mother–child body length correlation in A group mothers with the *second* child, as the environmental change would induce the child to achieve a genetic potential that the mother might not have ever realized. For weight and skinfolds, however, a concomitant change could occur as both mother and child take on body reserves, such that the correlation between A group mothers and their second children might be expected to increase over the B group correlations or the correlations between A group mothers and their first children.

RESULTS

In Table 2 are presented sibling correlations birth to 30 months of age. We have presented the correlations according to the sex of the treated or second infant, because sex seems to mediate the effectiveness of supplementation in this and other studies (Mora et al., 1979; McDonald et al., 1981; Mueller and Pollitt, 1982).

Sibling correlations in measurements at birth

Where the second or younger member of a pair was a male, sibling correlations for birthweight and head circumference at birth are highly significant ($p < 0.01$) in the placebo (B) group ($r = 0.56$ and 0.50 respectively) and are substantially lower in the supplemented (A) group ($r = 0.23$ and 0.25, respectively). The between-group difference in the correlations reaches the $p < 0.10$ level of statistical significance. This pattern is not evident where the younger member of a pair of siblings is a female. Sibling correlations are highly significant in both A and B groups ($r = 0.58$ and 0.61 for weight and head circumference, respectively, in the placebo group; the corresponding correlations for the supplement group are 0.61 and 0.60). To a lesser extent the sex-specific treatment pattern is evident in the other measurements (length, skinfold) except for the Rohrer Index. In both sex-specific groups the

TABLE 2. Sibling–sibling anthropometric correlations in Sui Lin, Taiwan, birth to 30 months, by sex of the second (supplemented in utero) sibling

Measurement	Sex of second infant/ supplement		Age (months)								
			0	1	3	6	9	12	18	24	30
Weight	Male	B	0.56	0.69	0.64	0.59	0.55	0.60	0.57	0.46	0.54
		A	0.23	0.25	0.28	0.49	0.38	0.43	0.40	0.44	0.43
	Female	B	0.58	0.62	0.48	0.38	0.33	0.33	0.38	0.33	0.26
		A	0.61	0.60	0.64	0.62	0.67	0.60	0.64	0.62	0.60
Length	Male	B	0.34	0.44	0.49	0.30	0.41	0.38	0.36	0.39	0.30
		A	0.21	0.30	0.40	0.31	0.28	0.30	0.30	0.40	0.35
	Female	B	0.46	0.57	0.54	0.43	0.46	0.37	0.43	0.42	0.31
		A	0.47	0.31	0.45	0.41	0.42	0.43	0.45	0.48	0.50
Head circumference	Male	B	0.50	0.41	0.34	0.34	0.32	0.39	0.33	0.37	0.34
		A	0.25	0.38	0.34	0.48	0.47	0.42	0.47	0.48	0.45
	Female	B	0.61	0.59	0.49	0.41	0.38	0.37	0.37	0.43	0.44
		A	0.60	0.43	0.46	0.34	0.39	0.45	0.41	0.50	0.50
Subscapular skinfold	Male	B	0.32	0.28	0.37	0.27	0.41	0.38	0.61	0.47	0.41
		A	(0.11)	0.24	0.34	0.37	0.29	0.24	0.41	0.41	0.54
	Female	B	(−0.07)	0.37	0.44	0.47	0.37	0.40	(0.22)	0.37	0.40
		A	(−0.07)	(0.04)	(0.18)	(0.10)	(0.13)	(−0.08)	(0.22)	0.33	0.42
Rohrer Index (wt/L^2)	Male	B	0.55	0.48	0.58	(0.23)	0.32	0.43	0.47	0.55	0.31
		A	(−0.02)	0.30	0.49	0.54	0.49	0.53	0.41	0.41	(0.16)
	Female	B	0.49	0.38	0.45	0.26	0.35	(0.18)	(0.22)	0.39	0.38
		A	(−0.02)	0.31	0.28	0.25	0.45	0.47	0.49	(−0.07)	0.41

B = placebo; A = supplement.
Sample sizes vary from 47 to 52 sibling pairs depending on the age group. Smaller sizes tend to occur beyond 18 months of age. Correlations in parentheses are *not* significant ($p > 0.05$).

pattern is striking: the B group siblings are significantly correlated in the Index (r's = 0.55 and 0.49) while the corresponding correlations in the A group are close to zero (−0.02), the between-group differences being statistically significant ($p < 0.01$).

Sibling correlations up to 30 months postbirth

Among males, reduced A group correlations appear to persist over the first 30 months of life for weight and skinfold. Among female children reduced A group correlations are present postnatally only for the skinfold. For the other measurements initial group differences appear to disappear by 3–6 months of age. In other words, initially low A group correlations climb rapidly as for example for male head circumference and Rohrer Index of both sexes. One curious trend is the tendency for placebo group (B) sibling correlations to fall from birth to 30 months, particularly for female children. These differences are statistically significant ($p < 0.01$) by 9 months of age.

Mother–child correlations

Mother child correlations birth to 30 months are presented in Table 3. Here we found no differences related to supplementation by sex of child (mother–son versus mother–daughter). Hence, sexes are combined in Table 3. The only significant ($p < 0.05$) between-treatment group difference at birth was the mother–child correlation for subscapular skinfold. However, the magnitude of the mother–child subscapular skinfold correlations among the A group sec-

TABLE 3. Mother–child anthropometric correlations in Sui-Lin, Taiwan, birth to 30 months

Measurement	Child/ supplement		Age (months)								
			0	1	3	6	9	12	18	24	30
Weight	First	B	0.39	0.26	(0.15)	(0.13)	0.21	0.22	0.17	0.25	0.23
		A	0.16	0.19	0.27	0.25	0.24	0.29	0.28	0.19	0.20
	Second	B	0.40	0.37	0.39	0.36	0.32	0.38	0.41	0.47	0.40
		A	0.30	0.29	0.26	0.27	0.37	0.42	0.38	0.35	0.30
Height	First	B	0.22	0.32	0.26	0.18	0.24	0.25	0.26	0.26	0.25
		A	(0.12)	0.04	0.18	0.22	0.25	0.25	0.24	0.19	0.23
	Second	B	0.17	0.19	0.25	0.25	0.31	0.29	0.32	0.33	0.34
		A	0.19	0.24	0.26	0.28	0.31	0.35	0.36	0.33	0.43
Triceps skinfold	First	B	0.28	(0.13)	0.16	0.16	0.32	0.22	(−0.08)	0.11	(−0.08)
		A	0.34	(0.14)	0.19	0.23	0.22	0.19	(0.01)	(0.09)	(0.03)
	Second	B	(0.07)	(0.00)	(0.09)	(0.14)	0.21	0.20	0.20	0.17	0.22
		A	0.22	(0.10)	0.17	(0.13)	0.17	0.33	(0.02)	(0.02)	(−0.12)
Subscapular skinfold	First	B	0.30	0.23	0.20	(0.12)	0.18	(0.07)	(0.09)	(0.15)	(0.09)
		A	0.29	(0.06)	(0.13)	(0.01)	(0.01)	0.23	(0.09)	0.28	0.00
	Second	B	(−0.01)	(−0.14)	(0.09)	(0.11)	(0.15)	(0.13)	(0.12)	0.21	(0.06)
		A	0.29	(0.12)	0.27	0.34	0.21	(0.13)	0.31	0.24	0.20

B = placebo; A = supplement.
Mother's weight and skinfold measurements are those taken at 1 month following the birth of the first or second child.
Correlations in parentheses are *not* significant ($p > 0.05$).
For all first children and for second children birth to 18 months, sample sizes range from 101 to 112 mother–child pairs. For second children at 24 and 30 months of age, sample sizes range from 74 to 94 pairs.

ond children ($r = 0.29$) is similar to the correlations in A and B groups with first child ($r = 0.29$ and 0.30, respectively). Thus, it is difficult to relate this effect to treatment. Nevertheless, in group A, the correlation between mothers and their second child for this skinfold is consistently greater than in group B at most ages birth to 30 months. This is presumed support for the hypothesis that there are concomitant changes in mother and child in central fat stores as a result of supplementation.

Mother–second child correlations for body weight are higher at most ages than mother–*first* child correlations *irrespective* of supplement group. Although statistical evaluation of the changes is complicated by the longitudinal nature of the data, the differences do seem substantial given the fact of a correlation between the data at times one and two.

DISCUSSION

A reduction in sibling covariation in the supplemented group (A) would suggest environmental covariation among siblings in anthropometry via nutritional supplementation of the mother. The pattern of sibling correlations in Rohrer's Index (wt/L^3) shows this expected pattern most strikingly. Sibling resemblance in relative weight (Rohrer's Index) for the supplement group, increases markedly from birth to 3 months (Table 2). We interpret this trend as due to a return to similarity of environments for the two siblings of the high-calorie supplement group as both siblings would have received supplement via lactation. Thus maternal supplementation may be expected to affect infant relative weight more than weight or length alone; and as Brandt (1979) has proposed, the Rohrer's Index (wt/L^3) is a sensitive indicator of in utero nutrition.

There is also a difference in the magnitude of the correlations between supplement treatment groups in birth weight and head circumference at birth (Table 2). However, for these measurements the effect depends on the sex of the supplemented child, males being more affected than females. Even though the sibling correlations in birthweight and head circumference did not differ significantly at an acceptable level (p = 0.10) between groups when the sex of the second child was male, B group correlations were of a magnitude seen in previous studies of the genetics of birthweight: from 0.41 to 0.62 in six studies of adjacent birth order siblings, with most near 0.5 (Robson, 1978). This agrees well with birthweight sibling correlations among children in the B group (r = 0.56 and 0.58 for male and female children, respectively) and A group children in which the supplemented child was a female (r = 0.60). Thus, the sibling birthweight correlation in A group when the second child was a male (r = 0.23) is unusually low and probably represents a real sex-mediated effect of supplementation on birth size. Moreover, elsewhere we have reported an intrafamilial supplement effect on mean birthweight of male but not female infants (McDonald et al., 1981).

Postnatally, sibling correlations become more similar overall between treatments. The exceptions seem to be the following: for male supplemented (A group) infants, weight correlations tend to remain depressed. For both sexes, subscapular skinfold sibling correlations remain lower in A than B group, especially in females. Possibly there are lasting effects of prenatal supplementation on central fat deposits.

A curious trend is the reduction in sibling correlation in weight among female placebo infants with time (Table 2, rows 3 and 4). In one well-nourished sample (Karn, 1956), sibling correlations in weight were high at birth (0.55), diminished at 6 months (0.41), but increased again by 2 years (0.59). The tendency for B group correlations to falter across time might reflect the increased vulnerability of body mass in nutritionally stressed groups. If so, supplementation appears to have had an effect on stabilizing growth, but there appears to be a differential effect on the timing of this intervention according to the sex of the infant. These comments are highly speculative but deserve further study.

Mother–child weight correlations in the United Kingdom (Tanner and Israelsohn, 1963) ran from 0.48 at 1 month to 0.47 by 2 years of age. Weight correlations tended to diminish somewhat from 3 to 9 months, a pattern evident also in the Sui Lin data (Table 3). Mother–child height correlations in the United Kingdom ran from 0.27 at 1 month to 0.48 by 2 years of age. This tendency for the mother–child height correlations to increase is also evident in the Sui Lin sample. Correlations between mothers and their second children in Sui Lin are more similar in magnitude to those reported by Tanner and Israelsohn (1963), than are the markedly lower mother–first child correlations. This seems true for both weight and height. The magnitude of the Sui Lin mother–child height correlations seem within the range, and exhibit similar time-related trends, as those reported for other well-nourished samples (Bayley, 1954; Garn and Rohmann, 1966; Tanner et al., 1970; Gerylovova and Bouchalova, 1974). For example, in the Fels study (United States), mother–child correlations begin at around zero and climb to 0.29 by 2 years of age (Garn and Rohmann, 1966). We speculate that the increased body weight correlation of mothers with their second children, may reflect improvements in the environment unconnected to supplementation. That Taiwan was undergoing dramatic improvements in countrywide health indicators during decades of which the study was a part, has been documented by Baumgartner (1982) and Gallin and Gallin (1982). Also Adair et al. (1982) have documented secular increase in maternal weight in Sui Lin over the 6.5 years spanned by the study.

There are a number of possible implications of the genetic approach we have used for assessing the effectiveness of maternal supplementation programs: (1) maternal supplementation may affect components of variation more than means of measurements. This is shown by the significant between-group differences in the sibling correlations for some anthropometrics but insignificant differences in mean birthweight between supplement groups (McDonald et al., 1981). (2) Growth of infant relative weight (Rohrer Index) may be more affected by prenatal supplementation than weight or length alone. Fetal length velocity is highest at midpregnancy; fetal weight velocity is highest in the last trimester of pregnancy (Tanner, 1978). The response in Rohrer's Index could indicate a greater effectiveness of maternal supplementation during the last trimester. This is also a time for rapid brain growth, hence, the response in head circumference and birth weight, although only in male fetuses. (3) The differential effects by sex of the infant, suggest that sex is an important mediator of supplementation, a result obtained in one other study of maternal nutritional supplementation (Mora et al., 1979). (4) Concomitant changes in central fat (subscapular) in mother and child may occur as a result of supplementation. Thus, growth in lean body components may be little affected by supplementation programs aimed solely at the mother.

ACKNOWLEDGMENTS

Dr. R. Quentin Blackwell was a principal coinvestigator of this study during the period of data collection. At that time he was staff member of the United States Medical Research Unit No. 2 in Taipei, Taiwan. His contribution in making this study possible is gratefully acknowledged.

We thank the University of Texas School of Public Health for provision of computer time and Gay Robertson and the Staff of the Word Processing Center for manuscript preparation. The Nestle Coordination Center for Nutrition and The Ford Foundation supported in part the data analysis for this study.

LITERATURE CITED

Adair, L, Pollitt, E, and Mueller, WH (1982) The Bacon Chow Study: Changes in maternal anthropometry during pregnancy and lactation. (submitted).

Baumgartner, R (1982) The Bacon Chow Study: The relationship between infectious illness and growth in infants. Ph. D. Dissertation. University of Texas Health Science Center, School of Public Health.

Bayley, N (1954) Some increasing parent–child similarities during the growth of children. J. Ed. Psychol. *45:*1–21.

Blackwell, RQ, Chow, BF, Chin, KSK, Blackwell, BN, and Hsu, SC (1973) Prospective maternal nutrition study in Taiwan: Rationale, study design, feasibility and preliminary findings. Nutr. Rep. Int. *7:*517–532.

Brandt, I (1979) Postnatal growth of preterm and full term infants. In FE Johnson, AF Roche, and C Susanne (eds): Human Physical Growth and Maturation, Methodologies and Factors. New York: Plenum Press, pp. 139–160.

Chow, BF (1974) Effect of maternal dietary protein on anthropometric and behavioral development of the offspring. In A Roche, and F Faulkner (eds): Nutrition and Malnutrition: Identification and Measurement. New York: Plenum Press, pp. 183–219.

Gallin, B, and Gallin, RS (1982) Socioeconomic life in rural Taiwan. Twenty years of development and change. Mod. China *8:*205–246.

Garn, SM, and Rohmann, CG (1966) Interaction of nutrition and genetics in the timing of growth and development. Pediatr. Clin. N. Am. *13:*353–379.

Gerylovova, A, and Bouchalova, M (1974) The relationship between children's and parents' heights in the age range 0–6 years. Ann Hum. Biol. *1:*229–232.

Herriott, RM, Hsueh, AM, and Aitchison, R (1977) Influence of maternal diet on offspring: Growth, behavior, feed efficiency and susceptibility (human). Final Report on AID/CSD 2944 Contract with the Johns Hopkins University, Baltimore.

Karn, MN (1956) Sibling correlations of weight and growth rate over the first five years. Ann. Hum. Genet. *21:*177–184.

McDonald, EC, Pollitt, E, Mueller, WH, Hsueh, AM, and Sherwin, R (1981) The Bacon Chow Study: Maternal nutritional supplementation and birthweight of offspring. Am. J. Clin. Nutr. *34:*2133–2144.

Martorell, R, Yarbrough, C, Lechtig, A, Delgado, H, and Klein, RE (1977) Genetic–environmental interaction in physical growth. Acta Pediatr. Scand. *66:*579–584.

Mora, JA, de Paredes, B, and Wanger, M (1979) Nutritional supplementation and the outcome of pregnancy. I. Birthweight. Am. J. Clin. Nutr. *32:*455–462.

Mueller, WH (1977) Sibling correlations in growth and adult morphology in a rural Colombian population. Ann. Hum. Biol. *4:*133–142.

Mueller, WH, and Pollitt, E (1982) The Bacon Chow Study: Effects of nutrition supplementation on sibling–sibling anthropometric correlations. Hum. Biol. *54:*455–468.

Mueller, WH, and Titcomb, M (1977) Genetic and environmental determinants of growth in a rural Colombian population. Ann. Hum. Biol. *4:*1–15.

Robson, EB (1978) The genetics of birthweight. In F Falkner and JM Tanner (eds): Human Growth. I. Principles and Pre-natal Growth. New York: Plenum Press, pp. 285–297.

Russell, M (1976) Parent–child and sibling–sibling correlations of height and weight in a rural Guatemalan population of preschool children. Hum. Biol. *48:*501–515.

Tanner, JM (1978) Foetus into Man: Physical Growth from Conception to Maturity. Cambridge, MA: Harvard University Press.

Tanner, JM, and Israelsohn, WJ (1963) Parent-child correlations for body measurements of children between the ages of one month and seven years. Ann. Hum. Genet. *26:*245–259.

Tanner, JM, Goldstein, H, and Whitehouse, R (1970) Standards for children's height ages 2–9 years allowing for height of parents. Arch. Dis. Child. *45:*755–762.

Wohlleb, JC, Pollitt, E, Mueller, WH, and Bigelow, R. The Bacon Chow Study: Maternal supplementation and post natal growth (in press).

AMERICAN JOURNAL OF PHYSICAL ANTHROPOLOGY 62:19–22 (1983)

Alpha-1-Antitrypsin-Deficient Phenotype is Not Maintained by Segregation Distortion

B.K. SUAREZ AND J.A. PIERCE
Departments of Psychiatry and Genetics, Washington University School of Medicine, and The Jewish Hospital of St. Louis, St. Louis, Missouri 63110 (B.K.S.), and Department of Internal Medicine, Pulmonary Disease Division, Washington University School of Medicine, and Department of Internal Medicine, John Cochran Veterans Administration Hospital, St. Louis, Missouri (J.A.P.)

KEY WORDS Segregation distortion, Alpha-1-antitrypsin, Pi^Z

ABSTRACT Recent reports have suggested that the alpha-1-antitrypsin allele Pi^Z, which in homozygotes results in severe deficiency of this important protease inhibitor, is maintained at a relatively high gene frequency through the mechanism of segregation distortion. We report here on 121 nuclear families selected because only one parent was segregating the Z allele. After correcting for ascertainment, no evidence of preferential transmission was observed in 278 informative offspring.

The plasma protein alpha-1 antitrypsin (α_1AT) is the most important broad spectrum protease inhibitor in man. Indeed, it is known as the *Pi* system because of its protease inhibiting capacity. As an acute phase reactive protein, produced in the parenchymal cells of the liver, its plasma levels covary positively with both inflammation and vaccination (Jacobsson, 1955; Kueppers, 1968).

The Pi system presents geneticists with a variety of unique and intriguing features, few of which are thoroughly understood. Taken collectively, these features suggest that the α_1AT system is subject to various selection pressures, some of which appear to act in opposite directions.

Included among α_1AT's more interesting features is its high level of polymorphism. With the recent widespread use of separator isoelectric focusing (IEF) techniques, the number of known public and private alleles has increased to approximately three dozen—all of which are codominantly expressed in heterozygotes. Along with the increased number of alleles that can be identified with IEF came an appreciation that a large segment of the population are Pi heterozygotes. Using traditional acid starch gel electrophoresis, for instance, only 10% of Europeans were thought to be heterozygotes (Fagerhol, 1972) whereas 30–40% can now be recognized (Frants and Eriksson, 1978).

A number of alleles at the Pi locus—notably Pi^Z, Pi^{null}, and to a lesser extent Pi^S—result in a reduction of circulating α_1AT. These same alleles are associated with a variety of disorders that, taken as an agregate, most likely confer a slight to moderate selective disadvantage on their carriers. In homozygotes the Z and null alleles are strongly associated with chronic obstructive pulmonary disease in adults (Laurell and Eriksson, 1963; Eriksson, 1978; Larsson, 1978) and cirrhosis of the liver in children (Sharp et al., 1969; Lieberman et al., 1972; Karitzky et al., 1977). The mean age of onset of emphysema in α_1AT-deficient persons is 38 for smokers and about 50 for nonsmokers (Larsson, 1978). Thus, while it is by no means apodictic that early onset pulmonary emphysema shortens the potential reproductive period (or otherwise interferes with parenting), the strength of the disease association makes it likely that the Z and null alleles may be subject to negative selection pressure. Nonetheless, they achieve a combined frequency that exceeds 1% in many Western populations (Pierce et al., 1975; Kueppers and Christopherson, 1978; Arnaúd et al., 1979; Klasen, 1980).

Recently three groups reported observing an excess of MZ offspring in the backcross mating ♂ MZ × ♀ MM (Chapuis-Cellier and Arnaúd,

Received June 5, 1982; accepted March 16, 1983.

1979; Iammarino et al., 1979; Frants and Eriksson, 1980). No excess of Pi^Z-bearing offspring were reported in backcross matings where the mother is the heterozygote. Even though these reports remain controversial (see, for instance, Cox, 1980; Mittman and Madison, 1980), the prospect of explaining the maintenance of the Z allele via segregation distortion represents an exciting development in human population genetics. We report here the results of a proband-based study designed to determine if indeed the Pi^Z allele is preferentially transmitted.

METHODS AND MATERIALS

Probands were Z heterozygotes or homozygotes through whom the extended family was ascertained. Some of these subjects originally were identified in the early 1970s when our facility was designated as one of two official U.S. Public Health Service α_1AT typing laboratories. Additional probands were identified from three sources; blood donors (Pierce et al., 1975), patients undergoing lung function tests, and patients referred by various physicians. Regardless of when a proband was originally ascertained, all were resampled so that IEF could be used to verify their phenotype.

After ascertaining the proband and Pi typing his or her offspring, a sequential strategy consisting of three sampling waves was followed wherever practical. In the primary wave the proband's parents and sibs were sampled. In the second wave the sibs of the informative parent (i.e., uncles and aunts) were sampled. Finally, in the third wave the offspring of informative sibs (i.e., nephew- and nieceships) and the offspring of informative uncles and aunts (i.e., cousinships) were α_1AT typed. This efficient sequential sampling strategy guaranteed that few uninformative nuclear families would be sampled. Simultaneously, the strategy increased the likelihood that, with the exception of the proband's immediate family, ascertainment for the kindred's remaining nuclear families would be through the parents thereby obviating the necessity for ascertainment corrections. Written informed consent was obtained from all participating family members.

Blood samples were obtained by venipuncture and sodium azide was added to a concentration of 0.02% w/v. Antitrypsin was quantitated by the rocket technique of Laurell (1972) using monospecific antiserum produced in our laboratory. Screening agarose gel electrophoresis was performed at pH 8.6 in 0.075 M barbital buffer, 1% w/v agarose (Johansson, 1972). Isoelectric focusing was performed with ampholine PAG plates (pH 4.0–5.0) obtained from LKB Aminkemi (Rockville, MD). Dithioerythritol was added to the samples to a concentration of 0.03 M and incubated at 37°C for 60 min prior to the isoelectric run (Pierce and Eradio, 1979).

RESULTS

Table 1 gives the results for informative matings segregating the Pi^Z allele. Mating types are reported separately by sex since the earlier reports suggested that segregation distortion occurs only when the father is heterozygous. The figures in parentheses are the number of probands. The matings involving an untyped parent (whose phenotype is denoted by "?") deserve special comment. Among these matings, those that failed to produce a Z bearing offspring (N = 2) are unambiguous with respect to the Pi^Z allele. Additionally, two matings that produced a mixture of offspring phenotypes were found to be unambiguous upon IEF subtyping of the family's M alleles. This leaves six ambiguous matings containing an untyped parent.

TABLE 1. Distribution of offspring in informative families segregating the Pi^Z allele

Mating type		Offspring	
father mother	Number of families	Received mother's Z	Did not receive mother's Z
MM × MZ	40	60 (9)	52
? × MZ	6	8 (1)	1
MS × MZ	4	6 (1)	5
MM × SZ	3	7 (1)	3
MP[1] × MZ	1	4	1
FM × MZ	1	1	1
ZZ × MZ	1	1	2
Total	56	87 (12)	65
		Received father's Z	Did not receive father's Z
MZ × MM	49	61 (8)	56
MZ × MS	4	6 (2)	3
MZ × ?	4	3	2
SZ × MM	2	1	1
SZ × MS	2	5	1
MZ × SS	1	1	1
FZ × MM	1	—	1
Z– × MM	1	3	1
MZ × FM	1	1 (1)	2
Total	65	81 (11)	68
Grand total	121	168 (23)	133

Figures in parentheses are the number of probands.
[1]P saintlouis.

Under the assumption that the six ambiguous matings are indeed backcrosses (with respect to the Z allele), and after removing all probands from the analysis, the segregation ratio is estimated as 75/140 = 53.5% (χ^2 = 0.71) when the mother is heterozygous and 70/138 = 50.7% (χ^2 = 0.03) when the father is heterozygous. Neither ratio deviates significantly from the value of 1/2 expected under the hypothesis of no segregation distortion. When the six ambiguous matings are deleted from the analysis the estimated segregation ratios are, respectively, 74/139 = 53.2% (χ^2 = 0.58) and 64/132 = 48.5% (χ^2 = 0.12) again neither of which differs statistically from 50%.

DISCUSSION

It is not obvious why some workers have found segregation distortion while others have not. One possibility is that families containing more than a single ZZ child may differentially be reported in the literature (Cox, 1980). This could arise, for instance, when a series of probands are ascertained through a hospital—especially one that specializes in children with cirrhosis—and ill sibs are preferentially Pi typed. Since ill sibs are more likely to be ZZ homozygotes, rather than heterozygotes or nondeficient homozygotes, an apparent segregation bias could result. While this possibility may help explain the results from intercross families, it fails to explain the apparent distortion observed by some workers for backcross families. Moreover, it fails to explain why preferential transmission has been observed only when the father is the Pi^Z-bearing heterozygote. For the data presented here, only a single family was capable of segregating a ZZ child and, as it happens, this family did not require an ascertainment correction since it was sampled through the parents.

Besides chronic obstructive pulmonary disease and childhood liver disease, a number of other disorders have been alleged to be associated with the Pi^Z allele. Indeed, a variety of diseases have been reported to be more frequent even in MZ heterozygotes. Included in this list are such diverse disorders as rheumatoid arthritis (Cox and Huber, 1976, 1980), ankylosing spondylitis (Buisseret et al., 1977), chronic pancreatitis (Novis et al., 1975), anterior uveitis (Brewerton et al., 1978), fibrosing alveolitis (Geddes et al., 1977), combined immune deficiency (Gelfand et al., 1979), periodontal disease (Peterson and Marsh, 1979), and psoriasis (Beckman et al., 1980). The majority of these associations were established on a small series of patients, usually less than 100, and show only a modest increase in relative risk. Attempts at replication have generally been unsuccessful (Benn andWood, 1976; Sjöblom and Wollheim, 1977; Brown et al., 1979; Karsh et al., 1979). Accordingly, with the exception of chronic obstructive pulmonary disease and childhood cirrhosis, it seems prudent to view these alleged associations as contentions until more compelling studies are conducted.

If segregation distortion is not responsible for the maintenance of alleles that give rise to α_1AT deficiency, are there any other mechanisms that could be responsible? Two mechanisms, both of which operate by increasing fertility, have been advanced as possible candidates. Fagerhol and Gedde-Dahl (1969) studied families selected for large sibship size and found an excess of matings involving the Pi^Z allele. These workers suggested that women with the Z allele may be more fertile than others and speculated that increased protease activity may enhance sperm migration, perhaps by decreasing the viscosity of cervical mucus. More recently Lieberman et al. (1979) have suggested that such an increase in fertility could be due to an increased rate of twinning. The preponderance of the data, however, clearly favors this explanation for the partially deficient Pi^S allele rather than the Z allele (Cook, 1975; and Martin, 1982). In this connection, we have recently searched our family material for evidence of segregation distortion for the partially deficient Pi^S allele (Suarez et al., 1982). From over 250 informative meioses we estimated a segregation ratio of 0.528 (134 Pi^S offspring: 120 not-Pi^S offspring, χ^2 = 0.77, $0.6 < p < 0.7$) after correcting for ascertainment. It is possible, of course, that very small perturbations in the segregation ratio favoring the Z allele (or the S allele) could escape detection except for sample sizes that are prohibitively large.

Recently Chakraborty et al. (1982) have argued that the frequency of the Pi^Z allele if maintained by segregation distortion from MZ males, would require heavy selection against ZZ homozygotes in order to be a stable polymorphism. From a survey of European populations they estimate selection coefficients on the order of 5–50%, which seems unreasonably large for natural populations. As an alternative, they speculate that maintenance of the Pi^Z allele may involve an interaction with other immunologic marker systems with uneven recombination values in the two sexes.

The evolutionary forces responsible for maintaining the Pi^Z allele at a relatively high frequency in many Western populations still need to be elucidated. It seems unlikely, however, that segregation distortion is among them.

ACKNOWLEDGMENTS

We wish to acknowledge the skillful laboratory assistance of Bibiana Eradio. We are indebted to F. Harlan, T. Przybeck, and R. Resta for their vigorous fieldwork. This study was supported, in part by United States Public Health Service grants GM 28067, MH 31302, MH 14677, and GM 28719 and by the Veterans Administration Medical Research Program 821 funds.

LITERATURE CITED

Arnaúd, P. Galbraith, RW, Faulk, WP, and Black, C (1979) Pi phenotypes of alpha$_1$-antitrypsin in Southern England: Identification of M subtypes and implications for genetic studies. Clin. Genet. *15:*406–410.

Beckman, G, Beckman, L, and Liden, S (1980) Association between psoriasis and the α_1-antitrypsin deficiency gene Z. Acta Derm. Venereol. (Stockh.) *60:*163–164.

Benn, RT, and Wood, PH (1976) Rheumatoid arthritis and alpha-1-antitrypsin. Lancet *ii:*147.

Brewerton, DA, Webley, M, Murphy, AH, and Ward, AM (1978) The α_1-antitrypsin phenotype MZ in acute anterior uveitis. Lancet *i:*1103.

Brown, WT, Mamelok, AE, and Bearn, AG (1979) Anterior uveitis and alpha-1-antitrypsin. Lancet *ii:*646.

Buisseret, PD, Pembrey, ME, and Lessof, MH (1977) α_1-Antitrypsin phenotypes in rheumatoid arthritis and ankylosing spondylitis. Lancet *ii:*1358–1359.

Chakraborty, R, Constans, J, and Majumder, PP (1982) Transmission of Pi-Z allele for α_1-antitrypsin deficiency: Population genetic considerations. Paper presented at the 33rd Annual Meeting, American Society of Human Genetics, Detroit, Michigan.

Chapuis-Cellier, C, and Arnaud, P (1979) Preferential transmission of the Z deficient allele of α_1-antitrypsin. Science *205:*407–408.

Clark, P, and Martin, NG (1982) An excess of the Pi^S allele in dizygotic twins and their mothers. Hum. Genet. *61:*171–174.

Cook, PJL (1975) The genetics of α_1-antitrypsin: A family study in England and Scotland. Ann. Hum. Genet. *38:*275–287.

Cox, DW (1980) Transmission of Z allele from heterozygotes for α_1-antitrypsin deficiency. Am. J. Hum. Genet. *32:*455–457.

Cox, DW, and Huber, O (1976) Rheumatoid arthritis and alpha-1-antitrypsin. Lancet *i:*1216–1217.

Cox, D, and Huber, O (1980) Association of severe rheumatoid arthritis with heterozygosity for α_1-antitrypsin deficiency. Clin. Genet. *17:*153–160.

Erikssoⁿ, S (1978) Proteases and protease inhibitors in chronic obstructive lung disease. Acta Med. Scand. *203:*449–455.

Fagerhol, MK (1972) The serum alpha-1-antitrypsin polymorphism. In Grouchy, De, Ebling, and Henderson (eds): Proceedings of the IVth International Congress on Human Genetics. Amsterdam: Excerpta Medica, pp. 277–285.

Fagerhol, MK, and Gedde-Dahl, T Jr. (1969) Genetics of the Pi serum types. Family studies of the inherited variants of serum alpha$_1$-antitrypsin. Hum. Hered. *19:*3–8.

Frants, RR, and Eriksson, AW (1978) Reliable classification of six Pi M subtypes by separator isoelectric focusing. Hum. Hered. *28:*201–209.

Frants, RR, and Eriksson, AW (1980) Pi M subtypes of α_1-antitrypsin in isolate studies. In AW Eriksson, HR Forsius, HR Nevanlinna, PL Workman, and RK Norio (eds): Population Structure and Genetic Disorders. London: Academic Press, pp. 199–210.

Geddes, DM, Webley, M, Brewerton, DA, Turton, CW, Turner-Warwick, M, Murphy, AH, and Ward, AM (1977) α_1-Antitrypsin phenotypes in fibrosing alveolitis and rheumatoid arthritis. Lancet *ii:*1049–1050.

Gelfand, EW, Cox, DW, Lin, MT, and Dosch, HM (1979) Severe combined immune-deficiency disease in patient with α_1-antitrypsin deficiency. Lancet *ii:*202.

Iammarino, RM, Wagener, DK, and Allen, RC (1979) Segregation distortion of the α_1-antitrypsin Pi Z allele. Am. J. Hum. Genet. *31:*508–517.

Jacobsson, K (1955) Studies on the trypsin and plasma inhibitors in human blood serum. Scand. J. Clin. Lab. Invest. 7(Suppl. 14):55–102.

Johansson, BG (1972) Agarose gel electrophoresis. Scand. J. Clin. Lab. Invest. 29(Suppl. 124):7–19.

Karitzky, D. Otto, M, and Martin, JP (1977) Ikterus prolongatus durch heterozygoten Mangel an Alpha-1-Antitrypsin? Klin. Wochenschr. *55:*1175–1176.

Karsh, J, Vergalla, J, and Jones, A (1979) Alpha-1-antitrypsin phenotypes in rheumatoid arthritis and systemic lupus erythematosus. Arthritis Rheum. *22:*111–113.

Klasen, EC (1980) Biochemical Genetics of Human α_1-Antitrypsin. Leiden: Elve/Labor Vincit.

Kueppers, F (1968) Gentically determined differences in the response of alpha$_1$-antitrypsin levels in human serum to typhoid vaccine. Humangenetik *6:*207–214.

Kueppers, F, and Christopherson, MJ (1978) Alpha$_1$-antitrypsin: Further genetic heterogeneity revealed by isoelectric focusing. Am. J. Hum. Genet. *30:*359–365.

Larsson, C (1978) Natural history and life expectancy in severe alpha$_1$-antitrypsin deficiency, Pi Z. Acta Med. Scand. *204:*345–351.

Laurell, CB (1972) Electroimmunoassay. Scand. J. Clin. Lab. Invest. *29*(Suppl. 124):21–37.

Laurell, CB, and Eriksson, S (1963) The electrophoretic α_1-globulin pattern of serum in α_1-antitrypsin deficiency. Scand. J. Clin. Lab. Invest. *15:*132–140.

Lieberman, J, Mittman, C, and Gordon, HW (1972) Alpha$_1$-antitrypsin in the livers of patients with emphysema. Science *175:*63–65.

Lieberman, J, Borhani, NO, and Feinleib, M (1979) α_1-Antitrypsin deficiency in twins and parents-of-twins. Clin. Genet. *15:*29–36.

Mittman, C, and Madison, R (1980) Transmission of Z allele from heterozygotes for α_1-antitrypsin deficiency. Additional supporting data. Am. J. Hum. Genet. *32:*457–458.

Novis, BH, Young, GO, Bank, S, and Marks, IN (1975) Chronic pancreatitis and alpha-1-antitrypsin. Lancet *ii:*748–749.

Peterson, RJ, and Marsh, CL (1979) The relationship of alpha$_1$-antitrypsin to inflammatory periodontal disease. J. Periodont. *50:*31–35.

Pierce, JA, and Eradio, BG (1979) Improved identification of antitrypsin phenotypes through isoelectric focusing in dithioerythritol. J. Lab. Clin. Med. *94:*826–831.

Pierce, JA, Eradio, BG, and Dew, TA (1975) Alpha-1-antitrypsin phenotypes in St. Louis. J. Am. Med. Assoc. *231:*609–612.

Sharp, HL, Bridges, RA, Krivit, W, and Freier, EF (1969) Cirrhosis associated with alpha-1-antitrypsin deficiency: A previously unrecognized inherited disorder. J. Lab. Clin. Med. *73:*934–939.

Sjöblom, KG, and Wollheim, FA (1977) Alpha-1-antitrypsin phenotypes and rheumatic diseases. Lancet *ii:*41–42.

Suarez, B, Pierce, JA, Resta, R, Harlan, F, and Reich, T (1982) Alpha-1-antitrypsin allele *Pi^S* fails to show segregation distortion. Hum. Hered. *32:*246–252.

AMERICAN JOURNAL OF PHYSICAL ANTHROPOLOGY 62:23–31 (1983)

Epidemiology and Genetics of Neural Tube Defects: An Application of the Utah Genealogical Data Base

L.B. JORDE, R.M. FINEMAN, AND R.A. MARTIN
Division of Medical Genetics, Department of Pediatrics, University of Utah Medical Center, Salt Lake City, Utah 84132

KEY WORDS Neural tube defects, Anencephaly, Spina bifida, Epidemiology, Genealogical index

ABSTRACT The distribution and prevalence of births with neural tube defects in Utah from 1940 to 1979 are analyzed with regard to prevalence rates, secondary sex ratios, seasonality, yearly rates, and time–space clustering. The overall prevalence rate of 1.00 per thousand live births is comparable to that of other populations in the western United States. Analysis of sex ratios indicates a substantially higher proportion of females than males. No significant secular trends or time–space clustering are observed. No seasonality is seen for spina bifida; however, the anencephaly cases are delivered more frequently in the early spring and fall months. Following linkage of the neural tube defect cases to the Utah Genealogical Data Base, application of the genealogical index method shows substantial familial clustering of the disease. The average inbreeding coefficient of the neural tube defect cases is not elevated over that of matched controls. The empirical recurrence risk for the disease is calculated to be 3%, and the heritability estimate is 70%. Likelihood analysis of pedigrees containing spina bifida occulta and spina bifida cystica indicates that they may segregate as an autosomal dominant trait with a penetrance of 75%.

Neural tube defects (NTDs), which include anencephaly, spina bifida, and encephalocele, are among the most common birth defects. The prevalence rate of NTDs varies from about 0.5 per 1,000 births in some oriental populations to 9 per 1,000 births in Belfast, Northern Ireland (Elwood and Elwood, 1980). Most anencephalics are stillborn; those that are born alive survive for no more than a few days. While the survival rate for individuals with spina bifida has increased with improved medical care, 20–50% die before reaching the age of 5 (Elwood and Elwood, 1980).

A large number of genetic and nongenetic mechanisms have been proposed and studied in an effort to determine the etiology of NTDs. Nongenetic factors that may be associated with NTD prevalence and distribution include diet, socioeconomic status, drug exposure, vitamin deficiency, maternal age, parity, season of birth, infections during pregnancy, and "overripeness" of egg cells (see Leck, 1974, 1977; and Elwood and Elwood, 1980, for reviews). It is well known that NTDs tend to cluster in families. To account for this, several genetic mechanisms have been considered, including a recessive gene (Book and Rayner, 1950; Fuhrmann et al., 1971), a dominant gene with reduced penetrance (Yen and MacMahon, 1968), a recessive x-linked gene (Toriello et al., 1980), cytoplasmic inheritance (Nance, 1969), and polygenic inheritance (Williamson, 1965; Lalouel et al., 1979; Pietrzyk, 1980). No clear relationship between any of these genetic or nongenetic factors and the genesis of NTDs has been established. The etiology of the disease remains essentially unknown.

In this study, we will summarize the results of our research in three areas: (1) epidemiologic studies of the prevalence and distribution of NTDs in Utah; (2) genealogical studies of familial clustering and recurrence risks; (3) likelihood analysis of modes of inheritance.

Received June 5, 1982; accepted March 16, 1983.

EPIDEMIOLOGY

We endeavored to ascertain all cases of NTDs born to Utah parents from 1940 through 1979. 979,873 birth certificates, 248,208 death certificates, 11,161 fetal death certificates, and records from Utah's major referral centers were examined. The NTD cases were divided into three major categories: (1) anencephaly (includes cranioschisis and craniorachischisis); (2) spina bifida (includes myelomeningocele, meningocele, and rachischisis; excludes spina bifida occulta); and (3) encephalocele (includes midline exencephaly, cranium bifidum, and encephalomeningocele). Table 1 lists the data sources for each major category of NTD. The sources are listed in the order in which they were accessed (i.e., the birth certificates were examined before the fetal death certificates, etc.). As expected, most cases of spina bifida and encephalocele were found on birth certificates, while most cases of anencephaly were found on fetal death certificates. Since NTDs are nearly always readily observable at birth, it is expected that virtually all cases should appear on either birth or fetal death certificates. In fact, Table 1 shows that 18% of the cases were not reported on these documents, indicating their inadequacy even for the enumeration of major, easily identified congenital malformations.

The overall prevalence at birth in Utah is 991 NTDs in 979,873 live births, or 1.01 per thousand live births. The rates for anencephaly, spina bifida, and encephalocele are 0.38, 0.56, and 0.07 respectively. There were 11,161 fetal deaths in Utah from 1940 to 1979 (Utah State Department of Health, 1981). If these are included in the denominator, the NTD prevalence rate becomes 1.00 per thousand. These rates are similar to those of other western U.S. populations (Elwood and Elwood, 1980) but lower than those of eastern U.S. populations (Milham, 1962; Naggan and MacMahon, 1967; Erickson, 1976). NTD rates tend to be quite high in England (Leck, 1977). It is interesting that in spite of the high percentage of English ancestry among Utah residents, the NTD rate is still quite low. This tends to corroborate the results of other studies (Naggan and MacMahon, 1967) in which the offspring of Irish immigrants to the United States had lower NTD rates than did their parents.

Among the 991 NTD cases ascertained, there were 62 that were associated with other findings or syndromes that were not secondary to the NTD (e.g., cleft lip and/or cleft palate, extrophy of the cloaca, etc.). Most of these cases probably represent etiologically distinct diseases. Thus, they are excluded from the results given below.

Table 2 gives the sex ratios found for each type of NTD. As in other reports, NTDs (especially anencephaly) are seen much more frequently in females than in males (Leck, 1977). Factors that may be responsible for this preponderance of females include differential prenatal survival of males and females (Polani, 1959; Bell and Gosden, 1978), differential sensitivity to gonadotrophin deficiency (Janerich, 1975), and x-linked genes in twin fetuses (Knox, 1970).

Seasonal variation in NTD rates may indicate the involvement of certain etiologic factors such as temperature and diet. Such variation has been found in some surveys (McKeown and Record, 1951; Elwood, 1970; Carter and Evans, 1973), but not in others (Milham, 1962; Frezal et al., 1964; Wehrung and Hay 1970; Flynt and Rachelefsky, 1973). Figure 1 presents the seasonal distribution of NTDs, and Figure 2 presents the seasonal distributions of anencephaly and spina bifida separately. A Kolmogorov–Smirnov one-sample test was used to determine whether these distributions differed significantly from a uniform distribution. While the total NTD and spina bifida distributions did not differ from the uniform, the anencephaly distribution did. The excess of anencephaly cases in the early spring and fall months, and the deficit in May, correspond closely to the distribution that Elwood (1975) found for Canadian populations. One complicating factor in the analysis of the anencephaly data is the fact that length of gestation period was not available. However, studies using date

TABLE 1. Sources of data

Source	Anencephaly	Spina bifida	Encephalocele	Total
Birth certificates	117	359	42	518(52.3%)
Fetal death certificates	239	50	7	296(29.8%)
Hospital records	1	78	15	94(9.5%)
Death certificates	17	60	6	83(8.4%)
Total	374	547	70	991

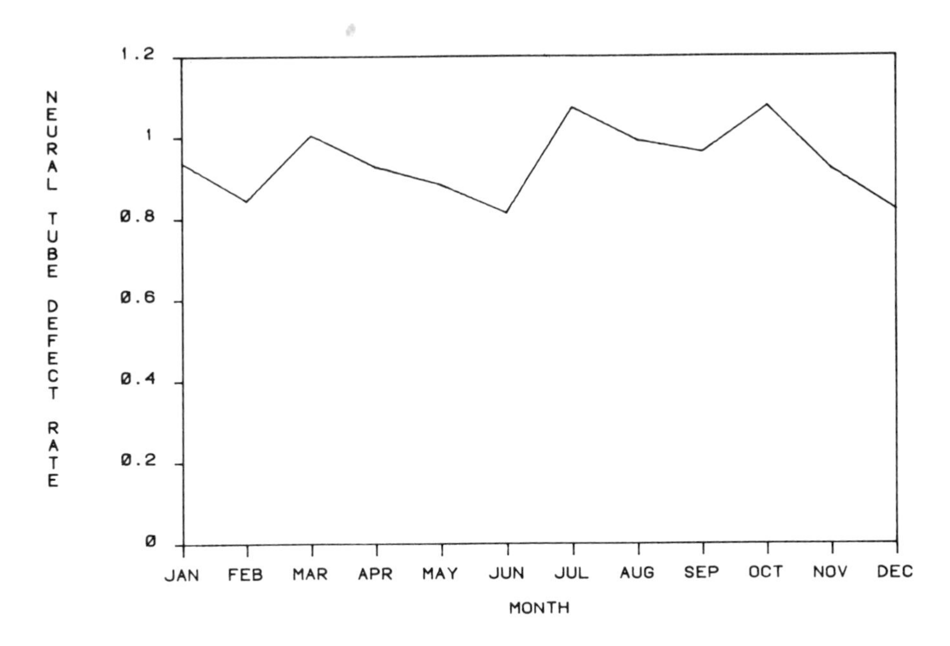

Fig. 1. Average monthly distribution of NTD rate (per thousand births).

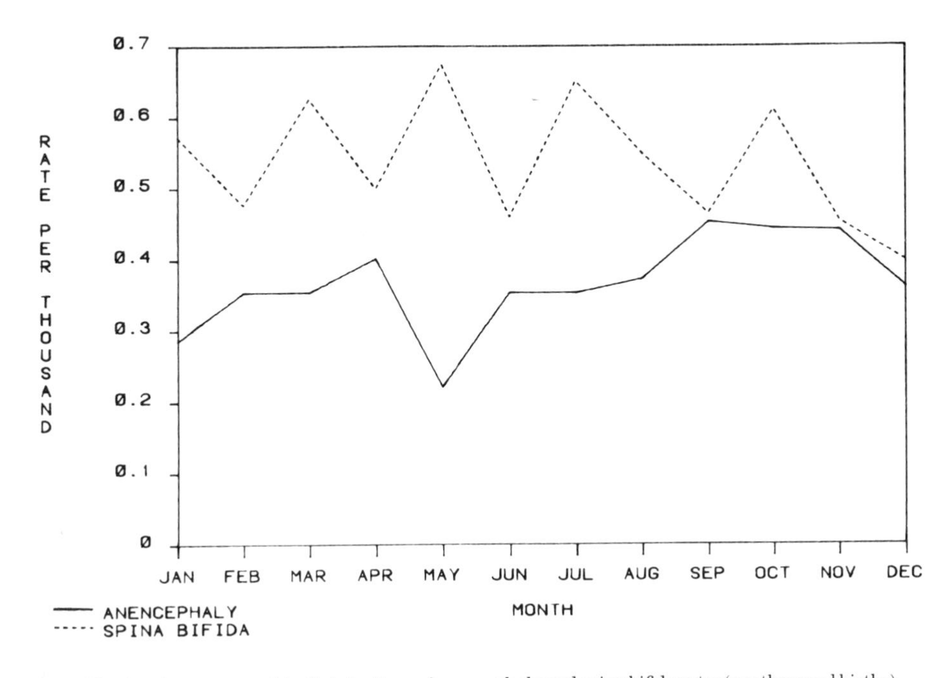

Fig. 2. Average monthly distributions of anencephaly and spina bifida rates (per thousand births).

TABLE 2. Sex ratios[1]

	Anencephaly	Spina bifida	Encephalocele	Total
Male	123	222	26	371
Female	230	285	36	551
Male ÷ female	0.53	0.78	0.72	0.67
Total	353	507	62	922

[1]Seven cases of unknown sex and 62 cases with associated malformations were omitted from this tabulation.

of conception and those using date of birth generally yield similar results (Elwood and Elwood, 1980).

Figure 3 shows the annual distribution of NTDs in Utah from 1940 through 1979. There are substantial year-to-year fluctuations in the data, and a linear regression analysis indicated no long-term trend. Since several other studies of U.S. populations have shown a long-term decline in NTD rates (MacMahon and Yen, 1971; Janerich, 1973; Windham and Edmonds, 1982), our result could indicate underreporting in the earlier years of the time frame.

To search of "epidemics" of NTDs in Utah (see, for example, Trichopoulos et al., 1971; Choi et al., 1972; Aylett et al., 1974), Knox's "all possible pairs" method (1963, 1964) was applied. The birth date of each case was used to calculate time differences (in days) between all possible pairs of cases, and the residence of the parents at the case's birth date was used to calculate spatial distances (in kilometers) between all possible pairs. 2 × 2 contingency tables were then formed (time distances vs. spatial distances), with various arbitrary cut-off levels used to denote "close" temporal and spatial clustering. Since the nature of a hypothesized "epidemic" was not known, a number of cut-off levels were tried: 1, 3, 5, 10, 20, 30, 50, and 100 km and 7, 15, 30, 60, 90, and 120 days. Since the pairs are not independently distributed, a chi-square test for significance would be inappropriate. Thus, following Knox (1964), the expected value of the upper-left cell of the contingency table was treated as the parameter of a Poisson distribution, and the probability of obtaining the observed value of the cell was estimated. No significant deviations from the expected values were seen for any of the cut-off levels in the spina bifida cases. Several cut-off levels did yield significant deviations ($0.02 < p < .05$) for the anencephaly cases. However, since 48 contingency tables were formed, the corrected significance level is 0.05/48, or approximately 0.001. Thus, these results were not actually significant after correction for multiple tests, and no time–space clustering can be inferred.

GENEALOGICAL STUDIES

To assess familial clustering and recurrence of NTDs, the ascertained cases were linked into the 1.2-million-member Utah Genealogical Data Base (see Skolnick, 1980, for a description of the data base). Two hundred and forty-nine (26%) of the NTD cases were found in the genealogical data base and thus were usable for the familial clustering analysis. There are two reasons why this figure is rather low. First, the individuals in the data base are nearly all members of the Mormon (Church of Jesus Christ of Latter-Day Saints) church, while many of the NTD families are not. Second, the data base tends to be more incomplete in later years. Because the controls are selected from the data base using a stratified random design (see below), and because only a *relative* comparison is made between cases and controls, we anticipate no important biases to result either from incomplete linkage or from incompleteness in the data base itself.

Familial clustering can be examined quantitatively using the "genealogical index" (Hill, 1980; Skolnick et al., 1981). The method consists of computing the coefficients of kinship (Malécot, 1969) between all possible pairs of cases. The resulting mean kinship value and the distribution of frequencies of kinship classes (e.g., sibs, first cousins, etc.) can be compared to the same values generated for sets of matched controls. These are selected randomly within each matching category from the genealogical data base. Since any number of control sets can be drawn, the mean kinship coefficients and the kinship distributions are averaged, and confidence intervals are computed. If the mean kinship coefficient for the cases lies outside the 95% confidence limits for the controls, there is evidence for familial clustering of the disease.

Fifteen control sets were run in this analysis. The controls were matched on birthplace (Utah vs. non-Utah—one non-Utah NTD case

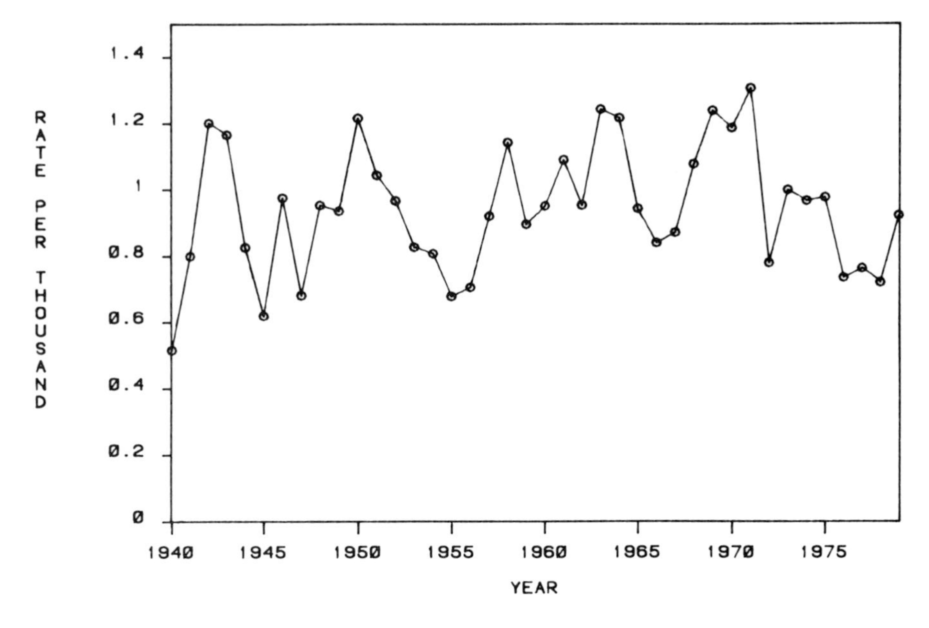

Fig. 3. Yearly frequencies of NTDs (per thousand births).

TABLE 3. Mean kinship and inbreeding coefficients (× 10^5) for NTD cases and controls

	Mean kinship coefficient	Confidence interval 95%	Mean inbreeding coefficient	Confidence interval 95%
Cases	10.00	—	6.28	—
Controls	1.56	1.33, 1.79	10.10	−2.95, 23.10

linked into the genealogy), birth date (5-year intervals), and sex. The results of this analysis are given in Table 3. The mean kinship coefficient of the cases is almost an order of magnitude higher than that of the controls, and it lies well outside of the range of the 95% confidence limits of the control sets. Thus, as expected, the NTD cases exhibit familial clustering. Figure 4 plots the number of individuals in each kinship class. A "kinship exponent" of 2, for example, represents the kinship coefficient of sibling pairs ($1/2^2$). Similarly a kinship exponent of 4, which denotes a coefficient of 1/16 ($1/2^4$), would most commonly signify first-cousin pairs. Most of the difference in kinship between cases and controls is due to 11 sib pairs who had NTDs. Familial clustering at this level could be due to both common environment and genetic effects. The NTD kinship is also elevated over that of the controls for coefficients of $1/2^6$, $1/2^7$, $1/2^8$, and $1/2^9$. At this level, common environment is much less likely to be a cause of familial clustering.

Table 3 also gives the mean inbreeding coefficients of cases and controls. Since consanguinity is quite low in this population (Woolf et al., 1956; Jorde, 1982), only one inbred individual was found among the cases (the product of a second-cousin marriage). The control sets typically included few if any inbred individuals. This resulted in a very wide confidence interval, which included the mean inbreeding coefficient of the cases within its boundaries. Thus, while consanguinity may play a role in the etiology of NTDs in some populations (Polman, 1951; Stevenson et al., 1966), there is no evidence that it does so in this population.

The empirical recurrence risk is the probability that a couple will produce a child with a certain disease, given that they have already produced one child with the disease. (Recurrence risks are sometimes also calculated for the case in which two affected children have already been born.) To evaluate empirical recurrence risks for NTDs, 198 families were analyzed. The number of sibs born after the birth of the first NTD child was 301. Of this number, nine had NTDs. This gives a recurrence risk of 2.99% (±1%), which is lower than the 5% figure given for British populations (Carter et al., 1968) but similar to that of other western American populations (McBride, 1979). Using Falconer's (1965) threshold model for polygenic traits, this recurrence risk and a prevalence rate of 1/1,000 yield a heritability estimate of approximately 70%. This is similar to the heritability values found in other populations (Carter, 1969; Carter and Evans, 1973; Pietrzyk, 1980). It is also similar to the value of 60% obtained in a Utah study of spina bifida patients (Woolf, 1975).

While the genealogical analyses cannot distinguish betweend common genes and common environment as causes for familial clustering, they do give some information regarding genetic mechanisms. The lack of any effect of consanguinity argues against a recessive mode of inheritance, although a strong consanguinity effect would not be expected for a trait as common as NTDs. In addition, the recurrence risks, which are well below the 25% and 50% figures expected for fully penetrant recessive and dominant genes, respectively, provide further evidence against a simple Mendelian genetic basis for NTDs. To gain further insight into the possible genetic causation of NTDs, likelihood analysis of specific pedigrees can be undertaken.

LIKELIHOOD ANALYSIS

Likelihood analyses of the segregation ratios of NTDs in families tend not to support a single-gene hypothesis (Lalouel et al., 1979, Pietrzyk, 1980). However, it has been proposed that spina bifida cystica, the typical "open-spine"

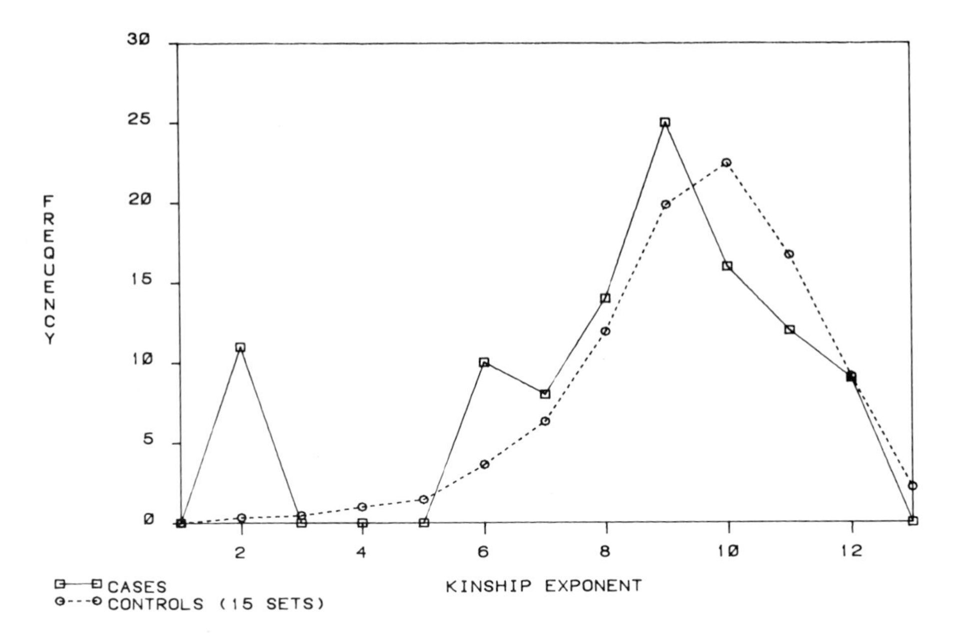

Fig. 4. Distribution of kinship of related pairs by kinship exponent. The control curve is the average of 15 sets of control groups.

TABLE 4. Penetrance estimates and log_{10} likelihood scores (disease frequency = 0.15)

Model	Penetrance			Log likelihood
	AA[1]	Aa	aa	
Sporadic	—	—	—	−26.69
Recessive	0	0	0.5838 ± 0.1042	−20.95
Intermediate	0.3613 ± 0.7764	0.8115 ± 0.1278	0	−18.80
Dominant	0.7492 ± 0.1002	0.7492 ± 0.1002	0	−18.90

[1]AA = homozygous dominant; Aa = heterozygote; aa = homozygous recessive.

condition, could represent the most severe expression of a gene which usually causes spina bifida occulta (Hindse-Nielsen, 1938; Sever, 1974; Fineman and Jorde, 1980), a relatively harmless spinal defect which is seen in 15–20% of the population. Family studies of the two forms indicate that they tend to be associated within families (Miller et al., 1962; Lorber and Levick, 1967; Laurence et al., 1971; Gardner et al., 1974; de Bruyere et al., 1977; Breslin and McCormack, 1979). Pedigree analyses (Mendell et al., 1974; Ruderman et al., 1977; Fellous et al., 1982) indicate that spina bifida occulta/cystica may be inherited as an autosomal dominant trait and may be loosely linked to the HLA complex and more tightly linked to PGM_3.

To examine further the inheritance patterns of spina bifida cystica and spina bifida occulta, we have begun pedigree analyses using the sequential sampling method to select pedigree members and correcting for ascertainment bias (Cannings and Thompson, 1977) (see Fineman et al., 1982, for details of the analysis). In four extended pedigrees, 63 individuals were x-rayed. Of these, 35 had spina bifida occulta or cystica, vertebral anomalies, and/or external defects. In the statistical analysis of these pedigrees, GEMINI (Lalouel, 1979) was used to estimate the penetrance parameters, and PAP (Hasstedt et al., 1979; Hasstedt and Cartwright, 1979) was used to calculate the $\log_{10}$ likelihoods of each model. Table 4 gives the $\log_{10}$ likelihoods for sporadic, recessive, intermediate, and dominant models. All three of the genetic models yielded much higher log likelihoods than the sporadic model. Among the genetic models, the $\log_{10}$ likelihood differences indicate that the intermediate and dominant models are each 100 times more "likely" than the recessive. Since the intermediate model has a large standard error associated with the homozygote penetrance parameter, and since it involves the estimation of two, rather than one, penetrance parameters, the dominant model is the most plausible. The penetrance for the spina bifida genotype is estimated to be 75%.

Like all statistical methods, likelihood analysis entails certain assumptions which are seldom fulfilled completely. These assumptions are too numerous and detailed to be dealt with here, but recent reviews are available (Conneally and Rivas, 1980; Elston, 1980; Morton, 1982). Because of these assumptions, and because of the explanatory weakness of the "dominant gene with reduced penetrance" result, our conclusions need to be strengthened substantially. One of the most effective ways to do this is to map the hypothesized gene for spina bifida to a specific chromosome. If linkage to a particular marker can be established, the gene can be followed in families, strengthening the evidence for a specific mode of inheritance (see Kravitz et al., 1979, for an example of this) and facilitating the separation of nongenetic from genetic expressions of the trait. To this end, we are currently enlarging our sample size and typing pedigree members for HLA, PGM_3, and other chromosome 6 markers (GLO, BF, and C4), as well as markers on other chromosomes.

ACKNOWLEDGMENTS

We are grateful for aid and discussion contributed by D.T. Bishop, J. Brockert, M. Dadone, S. Dintelman, S. Hasstedt, J. Gardner, T. Maness, and M. Skolnick. Financial support for this research was provided by grant number 6-291 from the March of Dimes Birth Defects Foundation.

LITERATURE CITED

Aylett, MJ, Roberts, CJ, and Lloyd, S (1974) Neural tube defects in a country town. Br. J. Prev. Soc. Med. *28:*177–179.

Bell, JE, and Gosden, CM (1978) Central nervous system abnormalities—contrasting patterns in early and late pregnancy. Clin. Genet. *13:*387–396.

Böök, JA, and Rayner, S (1950) A clinical and genetical study of anencephaly. Am. J. Hum. Genet. *2:*61–84.

Breslin, N, and McCormack, MK (1979) Risk factors associated with spina bifida. Am. J. Hum. Genet. *31:*69A (Abstr).

Cannings, C, and Thompson, EA (1977) Ascertainment in the sequential sampling of pedigrees. Clin. Genet. *12:*208–212.

Carter, CO (1969) Genetics of common disorders. Br. Med. Bull. *25:*52–57.

Carter, CO, and Evans, K (1973) Spina bifida and anencephalus in Greater London. J. Med. Genet. *10:*209–234.

Carter, CO, David, PA, and Laurence, KM (1968) A family study of major central nervous system malformations in South Wales. J. Med. Genet. *5:*81–106.

Choi, NW, Ateah, E, and Nelson, NA (1972) Some epidemiological aspects of central nervous system malformations in Canada. In MA Klingberg (ed): Drugs and Fetal Development. New York: Plenum Press, pp. 511–525.

Conneally, PM, and Rivas, ML (1980) Linkage analysis in man. In H Harris and K Hirschhorn (eds): Advances in Human Genetics, Vol 10. New York: Plenum Press, pp. 209–266.

de Bruyere, M, Kulakowski, S, Malchaire, J, Delire, M, and Sokal, G (1977) HLA gene and haplotype frequencies in spina bifida. Population and family studies. Tissue Antigens *10:*399–402.

Elston, RC (1980) Segregation analysis. In JH Mielke and MH Crawford (eds): Current Developments in Anthropological Genetics, Vol 1. New York: Plenum Press, pp. 327–354.

Elwood, JH (1970) Anencephalus in Belfast: Incidence and secular and seasonal variations, 1950–1966. Br. J. Prev. Soc. Med. *24:*78–88.

Elwood, JM (1975) Seasonal variation in anencephalus in Canada. Br. J. Prev. Soc. Med. *29:*22–26.

Elwood, JM, and Elwood, JH (1980) Epidemiology of Anencephalus and Spina Bifida. New York: Oxford University Press.

Erickson, JD (1976) Racial variations in the incidence of congenital malformations. Ann. Hum. Genet. *39:*315–320.

Falconer, DS (1965) The inheritance of liability to certain diseases, estimated from the incidence among relatives. Ann. Hum. Genet. *29:*51–76.

Fellous, M, Boue, J, Malbrunot, C, Wollman, E, Sasportes, M, Van Cong, N, Marcelli, A, Rebourcet, R, Hubert, C, Demenais, F, Elston, RC, Namboodiri, KK, and Kaplan, EB (1982) A five-generation family with sacral agenesis and spina bifida; possible similarities with the mouse t-locus. Am. J. Med. Genet. *12:*465–487.

Fineman, RM, and Jorde, LB (1980) Establishment of a birth defects registry with the aid of a genealogical data base. In J Cairns, JL Lyon, and MH Skolnick (eds): Banbury Report No. 4: Cancer Incidence in Defined Populations. Cold Spring Harbor, NY: Cold Spring Harbor Laboratory, pp. 319–331.

Fineman, RM, Jorde, LB, Martin, RA, Hasstedt, SJ, Wing, SD, and Walker, M (1982) The inheritance of spinal dysraphia as an autosomal dominant defect in four families. Am. J. Med. Genet. *12:*457–464.

Flynt, JW, and Rachelefsky, GS (1973) The epidemiology of anencephaly among blacks and whites in a southern United States community. In AG Motulsky and FJG Ebling (eds): 4th International Conference on Births Defects (Int. Congr. Ser. No. 297). Amsterdam: Excerpta Medica (Abstr), p. 95.

Frezal, J, Kelley, JK, Guillemot, ML, and Lamy, M (1964) Anencephaly in France. Am. J. Hum. Genet. *16:*336–350.

Fuhrmann, W, Seeger, W, and Bohm, R (1971) Apparently monogenic inheritance of anencephaly and spina bifida in a kindred. Humangenetik *13:*241–243.

Gardner, RJM, Alexander, C, and Veale, AMO (1974) Spina bifida occulta in the parents of offspring with neural tube defects. J. Genet. Hum. *22:*389–395.

Hasstedt, SJ, and Cartwright, PE (1979) PAP: Pedigree Analysis Package. Department of Medical Biophysics and Computing, University of Utah, Salt Lake City, Technical Report No. 13.

Hasstedt, SJ, Cartwright, PE, Skolnick, MH, and Bishop, DT (1979) PAP: A FORTRAN program for pedigree analysis. Am. J. Hum. Genet. *31:*135A (Abstr).

Hill, JR (1980) A survey of cancer sites by kinship in the Utah Mormon population. In J Cairns, JL Lyon, and MH Skolnick (eds): Banbury Report No. 4: Cancer Incidence in Defined Populations. Cold Spring Harbor, NY: Cold Spring Harbor Laboratory, pp. 299–318.

Hindse-Nielsen, S (1938) Spina bifida prognose: erblichkeit. Eine Klinische studie. Acta Chir. Scand. *80:*525–578.

Janerich, DT (1973) Epidemic waves in the prevalence of anencephaly and spina bifida in New York State. Teratology 8:253–256.

Janerich, DT (1975) Female excess in anencephaly and spina bifida: Possible gestational influences. Am. J. Epidemiol. *101:*70–76.

Jorde, LB (1982) Genetic structure of the Utah Mormons: Migration analysis. Hum. Biol. *54:*583–97.

Knox, EG (1963) Detection of low intensity epidemicity. Application to cleft lip and palate. Br. J. Prev. Soc. Med. *17:*121–127.

Knox, EG (1964) Epidemiology of childhood leukaemia in Northumberland and Durham. Br. J. Prev. Soc. Med. *18:*17–24.

Knox, EG (1970) Fetus–fetus interaction: A model aetiology for anencephalus. Dev. Med. Child. Neurol. *12:*167–177.

Kravitz, K, Skolnick, M, Cannings, C, Carmelli, D, Baty, B, Amos, B, Johnson, A, Mendell, N, Edwards, C, and Cartwright, G (1979) Genetic linkage between hereditary hemochromatosis and HLA. Am. J. Hum. Genet. *31:*601–619.

Lalouel, JM (1979) GEMINI—A computer program for optimization of a nonlinear function. Department of Medical Biophysics and Computing, University of Utah, Technical Report No. 14.

Lalouel, JM, Morton, NE, and Jackson, J (1979) Neural tube malformations: Complex segregation analysis and calculation of recurrence risks. J. Med. Genet. *16:*8–13.

Laurence, KM, Bligh, AS, Evans, KT, and Shurtleff, DB (1971) Vertebral abnormalities in parents and sibs of cases of spina bifida cystica, encephalocele, and anencephaly. Proceedings of the 13th International Congress of Pediatrics *5*(2):415–421.

Leck, I (1974) Causation of neural tube defects: Clues from epidemiology. Br. Med. Bull.*30:*158–163.

Leck, I (1977) Correlations of malformation frequency with environmental and genetic attributes in man. In JG Wilson and FC Fraser (eds): Handbook of Teratology, Vol 2. New York: Plenum Press, pp. 243–324.

Lorber, J, and Levick, K (1967) Spina bifida cystica: Incidence of spina bifida occulta in parents and in controls. Arch. Dis. Child. *42:*171–173.

MacMahon, B, and Yen, S (1971) Unrecognized epidemic of anencephaly and spina bifida. Lancet *I:*31–33.

Malécot, G (1969) The Mathematics of Heredity. San Francisco: WH Freeman.

McBride, ML (1979) Sib risks of anencephaly and spina bifida. Am. J. Med. Genet. *3:*377–387.

McKeown, T, and Record, RG (1951) Seasonal incidence of congenital malformations of the central nervous system. Lancet *I:*192–196.

Mendell, NR, Johnson, AH, Ruderman, RJ, Amos, DB, and Yunis, EJ (1974) Spina bifida and the HLA system: Evidence for linkage. Am. J. Hum. Genet. *26:*60A (Abstr).

Milham, S (1962) Increased incidence of anencephalus and spina bifida in siblings of affected cases. Science *138:*593–594.

Miller, JR, Fraser, FC, and MacEwan, DW (1962) The frequency of spina bifida occulta and rib anomalies in the parents of children with spina bifida aperta and meningocoele. Am. J. Hum. Genet. *14:*245–248.

Morton, NE (1982) Outline of Genetic Epidemiology. Basel: S Karger.

Naggan, L, and MacMahon, B (1967) Ethnic differences in the prevalence of anencephaly and spina bifida in Boston, Massachusetts. N. Engl. J. Med. *277:*1119–1123.

Nance, WE (1969) Anencephaly and spina bifida: A possible example of cytoplasmic inheritance in man. Nature *224:*373–374.

Pietrzyk, J (1980) Neural tube malformations: Complex segregation analysis and recurrence risk. Am. J. Med. Genet. *7:*293–300.

Polani, PE (1959) Nuclear chromatin of anencephalic foetuses. Lancet *II:*240–241.

Polman, A (1951) Anencephaly, spina bifida, and hydrocephaly. Genetica *25:*29–78.

Ruderman, RJ, Mendell, NR, Ruderman, JG, Johnson, AH, and Amos, DB (1977) Evidence for linkage between HLA and spinal malformation in man. Proceedings of the 5th International Conference on Birth Defects, p. 64 (Abstr).

Sever, LE (1974) A case of meningomyelocele in a kindred with multiple cases of spondylolisthesis and spina bifida occulta. Am. J. Med. Genet. *4:*94–96.

Skolnick, MH (1980) The Utah Genealogical Data Base: A resource for genetic epidemiology. In J Cairns, JL Lyon, and MH Skolnick (eds): Banbury Report 4: Cancer Incidence in Defined Populations. Cold Spring Harbor, NY: Cold Springs Harbor Laboratory, pp. 285–297.

Skolnick, M, Bishop, DT, Carmelli, D, Gardner, E, Haldley, R, Hasstedt, S, Hill, JR, Hunt, S, Lyon, JL, Smart, CR, and Williams, RR (1981) A population-based assessment of familial cancer risk in Utah Mormon genealogies. In FE Arrighi, PN Rao, and E Stubblefield (eds): Genes, Chromosomes, and Neoplasia. New York: Raven Press, pp. 477–500.

Stevenson, AC, Johnston, HA, Stewart, MIP, and Golding, DR (1966) Congenital malformations: A report of a study of series of consecutive births in 24 centers. Bull WHO 34:1–125.

Toriello, HV, Warren, ST, and Lindstrom, JA (1980) Possible x-linked anencephaly and spina bifida—report of a kindred. Am. J. Med. Genet. *6:*119–121.

Trichopoulos, D, Desmond, L, and MacMahon, B (1971) A study of time–space clustering in anencephaly and spina bifida. Am. J. Epidemiol. *94:*26–30.

Utah State Department of Health (1981) Utah Vital Statistics Annual Report: 1979.

Wehrung, DA, and Hay, S (1970) A study of seasonal incidence of congenital malformations in the United States. Br. J. Prev. Soc. Med. *24:*24–32.

Williamson, EM (1965) Incidence and family aggregation of major congenital malformation of central nervous system. J. Med. Genet. *2:*161–172.

Windham, GC, and Edmonds, LD (1982) Current trends in the incidence of neural tube defects. Pediatrics *70:*333–337.

Woolf, CM (1975) A genetic study of spina bifida cystica in Utah. Soc. Biol. *22:*216–220.

Woolf, CM, Stephens, FE, Mulaik, DD, and Gilbert, RE (1956) An investigation of the frequency of consanguineous marriages among the Mormons and their relatives in the United States. Am. J. Hum. Genet. *8:*236–252.

Yen, S, and MacMahon, B (1968) Genetics of anencephaly and spina bifia? Lancet *II:*623–626.

AMERICAN JOURNAL OF PHYSICAL ANTHROPOLOGY 62:33–49 (1983)

Cultural and Biological Inheritance of Plasma Lipids

D. C. RAO, W. R. WILLIAMS, M. McGUE, N. E. MORTON, C. L. GULBRANDSEN, G. G. RHOADS, A. KAGAN, P. LASKARZEWSKI, C. J. GLUECK, AND J. M. RUSSELL
Division of Biostatistics, Department of Preventive Medicine, Washington University, St. Louis, Missouri 63110 (D.C.R., M.M., J.M.R.); Department of Genetics, The University of Texas System Cancer Center, M. D. Anderson Hospital and Tumor Institute, Texas Medical Center, Houston, Texas 77030 (W.R.W.); Population Genetics Laboratory (N.E.M.), John A. Burns School of Medicine (C.L.G.) and School of Public Health (G.G.R., A.K.), University of Hawaii, Honolulu, Hawaii 96822; The Lipid Research Clinic, The General Clinical Research Center, The CLINFO Center, University of Cincinnati, College of Medicine, Cincinnati, Ohio 45267 (P.L., C.J.G.)

KEY WORDS Cholesterol, Triglyceride, Plasma Lipids, Path analysis, Between-study-heterogeneity, Familial resemblance, Bivariate analysis

ABSTRACT A path analytic model for the analysis of nuclear family data is described and used to analyze the results of two major studies of cholesterol (CH) and triglyceride (TG), the Honolulu Heart Study (HHS) of Japanese–Americans and the Cincinnati Lipid Research Clinic (LRC) study of Caucasians. The studies were first analyzed separately to assess evidence for genetic and cultural transmission, marital resemblance, and maternal environmental effects for the two plasma lipids, and then simultaneously to identify the nature and sources of any between-study-heterogeneity. There were significant sources of heterogeneity between the two studies for CH (only marital environmental resemblance and non-transmitted sibling environmental resemblance) and for TG (only non-transmitted sibling environmental resemblance). The two studies were homogeneous with respect to the magnitude of genetic and cultural effects; for CH genetic heritability was estimated as $h^2 = .594 \pm .041$ while cultural heritability was estimated as $c^2 = .035 \pm .008$, and for TG the two heritabilities were estimated as $h^2 = .259 \pm .034$ and $c^2 = .108 \pm .014$. An additional bivariate analysis of the association between the two lipids revealed that all phenotypic resemblance could be explained in terms of an association of non-transmitted residual environments with little evidence for a genetic association. The relevance of these results for an understanding of the genetic epidemiology of plasma lipids is discussed.

Plasma lipids are important risk factors for coronary heart disease, and as such an understanding of the causes of family resemblance for plasma lipids is of primary importance. Clear understanding of these causes should facilitate identification of individual characteristics of the phenotype associated with genetic and environmental factors, which, in turn, should improve both the efficiency and cost effectiveness of primary prevention of atherosclerosis.

Familial aggregations of lipids and lipoproteins have been assessed in different studies based on different populations. Some of these studies were limited to plasma lipids alone, while others investigated lipoprotein concentrations as well. Here we shall investigate the multifactorial basis of the two plasma lipids, total cholesterol (CH) and triglyceride (TG), based on the following two studies: (1) *HHS Study:* the Honolulu Heart Study of Japanese–American families living on the island of Oahu (Rhoads et al., 1976), and (2) *LRC Study:* the Cincinnati Lipid Research Clinic Princeton School district family study of Caucasians in and around Cincinnati (Morrison et al., 1982). The analysis of the HHS study presented here

Received June 10, 1982; accepted March 16, 1983.

is new, while an analysis of the Cincinnati LRC has been recently presented elsewhere (Rao et al., 1982a). Analyses reported here are based on the models and methods of path analysis developed for the analysis of family data on quantitative traits. After briefly reviewing the path model and the statistical method of analysis, we shall present the results separately for each of the two studies. We shall then explore the possible heterogeneity between the studies, followed by a bivariate analysis of the two lipids.

A LINEAR MODEL

Analyses presented here are based on the fundamental linear structural equation given by

$$P = hG + cC + rR$$

where P = phenotype, G = genotype, C = family environment (also referred to as indexed environment, or familial environment), R = residual, and h, c, and r are standardized partial regression coefficients or *path coefficients.* All variables are assumed to be standardized. For the purposes of this article, G, C, and R are assumed to be uncorrelated. Therefore, taking variances on both sides of the structural equation yields

$$1 = h^2 + c^2 + r^2,$$

which is called an *equation for complete determination,* and gives rise to an inequality constraint on the two basic parameters of the model:

$$1 - h^2 - c^2 \geq 0,$$

where h^2 = genetic heritability, and c^2 = cultural heritability.

As family environment is not directly measurable, we introduced an *environmental index* (I) as an estimate of the unknown family environment, C (Morton, 1974; Rao et al., 1974). An index is not assumed to be a perfect estimator of C: a path coefficient from C to I measures the precision of I as an estimate of C. For quantitative traits like plasma lipids an index is constructed for each member of a family through stepwise multiple regression of the phenotype (P) on a number of relevant variables as described later. In this way we generate two "observed" variables for each individual, a phenotype (P) and an index (I). Figure 1 presents a path model for nuclear families that incorporates cultural transmission and marital resemblance, the latter representing the combined effects of assortative mating and cohabitation. The ten unknown parameters of the model are defined in Table 1. Marital resemblance is represented by correlated environments of spouses (u). Intergenerational differences in heritabilities are permitted: whereas h^2 and c^2 are the genetic and cultural heritabilities in children, they are h^2z^2 and c^2y^2 in adults. Specific maternal effects are included by distinguishing the effects of paternal (f_F) and maternal (f_M) environments on that of a child they rear. Separate indexed environments and indices are allowed for the children. A child's indexed environment is determined partly by parental environments, partly by a nontransmitted common sibship environment (B), and partly by residual causes. Finally, the path coefficient from indexed environment to index is i for young children and iv for adults, both measuring the extent to which indices are adequate estimates of the indexed environments (Rao et al., 1979a). Expected correlations between pairs of variables, such as father's phenotype and child's index, are derived from Figure 1 following the basic rules of path analysis (e.g., Li, 1975).

STATISTICAL ANALYSIS

Family data on quantitative traits such as serum lipids, after suitable adjustments to remove age and sex effects, can be analyzed by one of three methods (Rao et al., 1979b, 1983). In the first method, the data are first summarized in terms of estimates of familial correlations to which a linear model is then fitted assuming the correlation estimates to be independent. The second method also begins by summarizing the data in terms of estimates of the familial correlations, but now the correlations between these correlation estimates are empirically incorporated into the analysis when fitting path models. In the third method, the linear model is fitted directly to the quantitative data in much the same way as Lange et al. (1976) used in their variance components model. All methods assume multivariate normality of the data. Analyses presented here were performed using the first method, which is briefly described below.

Consider a set of sample correlation coefficients $r_1, r_2, \ldots, r_m$, estimated from family data, and their respective sample sizes $n_1, n_2, \ldots n_m$. We convert a sample correlation coefficient into Fisher's z transformation (Fisher, 1921) given by

$$z_i = 1/2 \ln \frac{1 + r_i}{1 - r_i}$$

which is asymptotically normally distributed

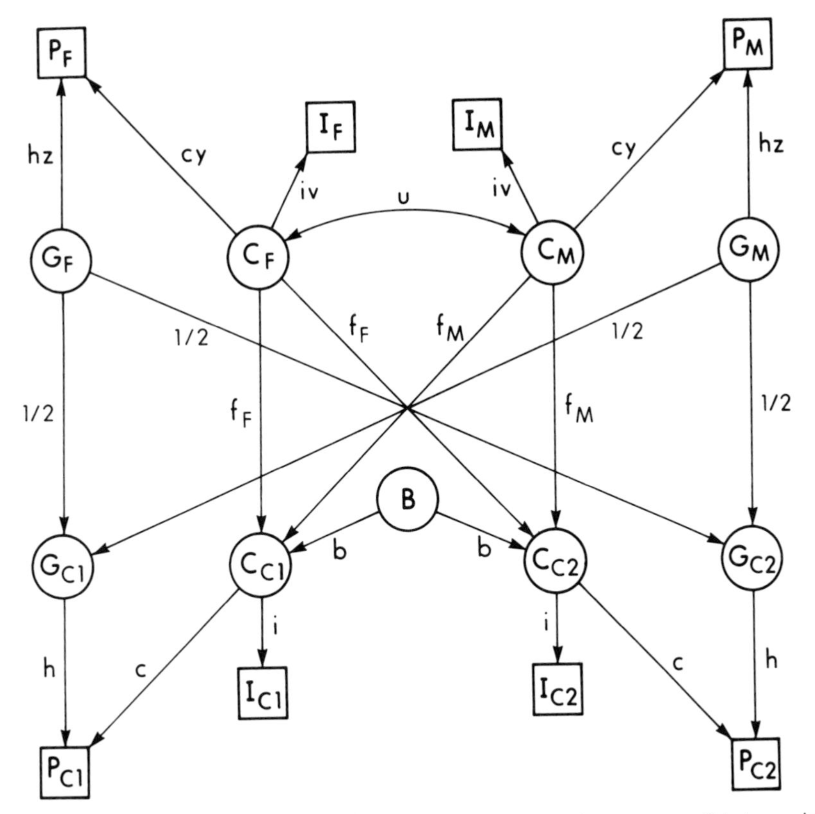

Fig. 1. Path diagram showing cultural and biological inheritance in nuclear families. The subscripts F, M, C1, and C2 denote father, mother and two children respectively. P is phenotype, G is genotype, C is transmissible environment with Index I, and B is nontransmitted common sibship environment.

TABLE 1. Definition of the ten parameters of the model (Figure 1)

Parameter	Definition
h	Effect of genotype on child's phenotype (square root of genetic heritability)
hz	Effect of genotype on adult's phenotype
c	Effect of environment on child's phenotype (square root of cultural heritability)
cy	Effect of environment on adult's phenotype
u	Correlation between parental environments
b	Effect of non-transmitted common sibship environment on child's environment
f_F	Effect of father's environment on that of a child he rears
f_M	Effect of mother's environment on that of a child she rears
i	Effect of environment on child's index
iv	Effect of environment on adult's index

with mean

$$\bar{z}_i = 1/2 \ln \frac{1 + \rho_i}{1 - \rho_i}$$

and approximate variance, $V(z_i) = 1/n_i$, where ρ_i is an expected correlation derived under a given model. Assuming asymptotic normality and independence of the z's, the overall log-likelihood function is approximated by

$$\ln L = -\chi^2/2 + \text{constant}$$

where

$$\chi^2 = \sum_{i=1}^{m} n_i(z_i - \bar{z}_i)^2.$$

We can estimate some or all of the unknown parameters by minimizing χ^2, or equivalently maximizing the log-likelihood. When k parameters are estimated, the residual χ^2 asymptotically follows a chi-square distribution with m − k degrees of freedom, which provides a goodness-of-fit test for the model. The residual χ^2 can also be used to carry out likelihood ratio tests of specific null hypotheses. For example, let χ^2_{m-k-w} be the residual χ^2 after estimating k + w parameters, and χ^2_{m-k} be another value after estimating only k of the k + w parameters, the other w parameters being fixed under a null hypothesis. Then $\chi^2_w = \chi^2_{m-k} - \chi^2_{m-k-w}$ is asymptotically distributed as a chi-square with w degrees of freedom (df) under the null hypothesis, providing the likelihood ratio test of the null hypothesis. In practice, parameters are estimated and hypotheses tested by minimizing χ^2 subject to a set of nonlinear inequality constraints on the parameters, such as the one described before, as discussed in Morton et al. (1983).

These methods have been implemented in PATHMIX, a FORTRAN computer program written for HARRIS computers (Morton et al., 1983; Rao et al., 1982a). In the first method, PATHMIX estimates 16 sample correlations with distinct expectations from nuclear family data, and then fits the path model to the sample correlations.

We note that a fundamental assumption of this method of analysis is that the different sample correlation coefficients are independent. Clearly, estimates of correlations obtained from the same set of nuclear families are not all independent of one another (Elston, 1975; Rao et al., 1979b). However, several analyses performed under independence and nonindependence assumptions gave essentially similar results (Gulbrandsen et al., 1979; Krieger et al., 1980; Morton et al., 1980; Rao et al., 1979b), which justifies the first method. Furthermore, recent evidence suggests that all three methods of analysis (i.e., assuming independent correlation estimates, incorporating correlations between correlations, and fitting path models directly to the family data) give similar results (Rao et al., 1983).

PLASMA LIPIDS

Analyses of the two studies are now presented separately.

The Honolulu heart study

From 1970–1972 a probability sample was drawn from the Honolulu Heart Study cohort of Japanese males born during the 20 years before 1920 and living on the island of Oahu in the State of Hawaii. From the remainder of the cohort, a selected sample was chosen on the basis of the upper 10% of casual cholesterol, the upper 10% of casual triglyceride, definite coronary heart disease, and/or definite cerebrovascular disease. Individuals were excluded who refused participation, who were nonfasting at the time of examination, or whose lipoprotein typing was incomplete. Data on the 2780 participants have been reported (Rhoads et al., 1976). A control sample was randomly selected from the above probability sample to investigate family resemblance for lipids and lipoproteins.

From the remainder of the subjects (those not in the probability sample), a selected sample was defined on elevated fasting beta-cholesterol or triglyceride or myocardial infarction confirmed by ECG. Individuals without a living wife and children were excluded from both samples. The 496 men and their 1850 wives and children constitute the Honolulu Heart Study family study, HHS in short (Gulbrandsen et al., 1977). Three aspects of this study need to be remembered. First, the lipid levels of wives and children were determined a few years after those of the fathers. Second, the families were ascertained largely through fathers with elevated lipid levels. Third, as a result of selection through elderly fathers, most children were adults and living separately from their parents and sibs.

These data were first adjusted for the effects of age and sex through stepwise multiple regression. In fact, each lipid variable was separately regressed on sex, age, age^2, age^3, sex × age, sex × age^2, sex × age^3, obesity as defined by Quetelet index (weight/$height^2$), hematocrit (percentage), smoking (number of cigarettes per day), and alcohol consumption (number of bottles of beer and glasses of wine or liquor per week), retaining only the significant terms (Gulbrandsen et al., 1977):

$$X = f_1(A,S) + g_1(S) + e_1$$

where X = lipid variable (CH or TG), $f_1(A,S)$ = polynomial in age and sex, $g_1(Z)$ = linear function of obesity, hematocrit, smoking, and alcohol and e_1 = residual. After fitting this equation, age–sex adjusted lipid value (phenotype, P), and an estimate of the familial environment (environmental index, I), were defined as

$$P = X - \hat{f}_1(A,S)$$

and

$$I = \hat{g}_1(Z),$$

where $\hat{f}_2$ and $\hat{g}_1$ denote the estimated contributions.

The assumptions which underly construction of indices are sometimes misunderstood and warrant some comment here. The first question that often arises is which variables should be included in the construction of the index. Ideally, we prefer to include in the above regression ($g_1(Z)$) only relevant environmental variables. In which case, the higher the correlation of the index variables with the phenotype the more informative the index will be. In practice, we do not always have many relevant variables or we may not be sure of the genetic basis of these variables. For example in the present case obesity (as measured by Quetelet index) is used in the construction of the environmental indices for the lipids. In a recent review, Garn et al. (1980) concluded that obese individuals have obese relatives whether they are genetically related or not, suggesting that the familial aggregation of obesity is primarily due to common environmental factors rather than common genes. If this is true, then obesity is an ideal index variable. If not true, then it does appear that the genetic component to obesity is small at best (Glueck et al., 1983). In the latter case, the important assumption is that the genetic factors of the index variables, should they exist, are not correlated with the genetic factors of the lipids. If such a correlation does exist, we would expect our analysis to yield an underestimation of the genetic heritability (h^2) and an overestimation of the cultural heritability (c^2) (see Rao et al., 1983). However, so long as the genotype of the lipid and the genotype of the index are uncorrelated, these heritabilities will not be altered (even if the index variables, such as obesity, are partly genetically determined).

Family data on P and I were initially analyzed using an earlier version of a path model (Rao et al., 1979b), which assumed one family environment for all children in a family, and therefore, one index for each sibship. Here we reanalyze the same family data according to the more general model presented here (Fig. 1). The 16 correlation coefficients were estimated for both CH and TG using PATHMIX and are presented in Table 2 together with estimated sample sizes (for details see Morton et al., 1983; Rao et al., 1982a). The mode of ascertainment was ignored when estimating the correlations. This, coupled with the prior testing of fathers, should reduce the father–child correlation (relative to other correlations, notably the mother–child correlation). For this reason, and because of previous experience with these data (Rao et al., 1979b), we analyzed the data two ways: once by adding a new parameter, t, that denoted a direct correlation between father's and child's phenotypes, and once by rejecting the father–child phenotypic correlation (and $t = 0$). In the first case, we expect

TABLE 2. Maximum likelihood estimates of familial correlations and their sample sizes for plasma lipids, based on data from the Honolulu Heart Study (HHS) and the Princeton School District family study (Cincinnati LRC)

Pair of variables	Honolulu heart study				Cincinnati LRC study[1]			
	Cholesterol		Triglyceride		Total cholesterol		Triglyceride	
	r	n	r	n	r	n	r	n
P_F, P_M	−0.070	393	−0.008	438	−0.016	88	−0.025	98
I_F, I_M	−0.110	425	0.057	399	0.178	89	0.180	83
(P_F, I_M) or (P_M, I_F)	−0.012	822	0.004	827	0.054	177	0.095	179
(P_F, I_F) or (P_M, I_M)	0.147	934	0.291	946	0.234	230	0.396	212
P_{C1}, P_{C2}	0.286	749	0.140	1068	0.380	176	0.200	170
I_{C1}, I_{C2}	0.168	932	0.140	1132	0.477	177	0.502	148
P_C, I_C	0.187	1202	0.319	1301	0.236	258	0.394	283
P_{C2}, I_{C1}	−0.006	1485	0.068	1918	0.152	264	0.179	270
P_F, P_C	0.186	818	0.144	985	0.362	146	0.204	137
I_F, P_C	0.034	841	0.052	886	0.058	110	0.044	125
P_F, I_C	−0.052	901	0.052	979	0.110	112	0.076	100
I_F, I_C	0.023	985	0.139	912	0.158	108	0.164	106
P_M, P_C	0.304	791	0.165	866	0.249	135	0.047	175
I_M, P_C	0.032	735	−0.021	837	0.097	137	0.047	173
P_M, I_C	0.008	814	0.027	884	−0.080	135	−0.004	146
I_M, I_C	0.218	933	0.164	930	0.261	158	0.268	161

[1]This study had data on 13 pairs of half-sibs also. See footnote to Table 5.

a negative estimate of t to compensate for the possible underestimation of the father–child correlation.

Both analyses of total cholesterol (CH) are summarized in Table 3. As previously found, t is significantly negative ($\chi_1^2 = 13.64 - 7.77 = 5.87, p < 0.02, t = -0.113$). Except for this, both analyses yielded identical results. The general model provided a good fit ($\chi_5^2 = 7.77$, $p > 0.16$). As before (Rao et al., 1979b), the combined effect of assortative mating and cohabitation is significantly negative ($\chi_5^2 = 13.18 - 7.77 = 5.41$, $p < 0.02$). The nontransmitted common sibship environmental effect is not significant ($\chi_1^2 = 11.38 - 7.77 = 3.61$, $p > 0.05$), but maternal effects are ($\chi_1^2 = 27.27 - 7.77 = 19.50$, $p < 0.0001$). The combined hypothesis with no intergenerational differences and no sibship environmental effect ($y = z = 1$, $b = 0$), fits very well ($\chi_8^2 = 11.44$, $p > 0.17$). However, some of the parameter estimates are considerably different from those under the general model, and therefore we take $y = z = 1$ as the most parsimonious hypothesis. It may be noted that our previous model and analysis (Rao et al., 1979b) imply, in terms of the present model, the following equality constraint:

$$b^2 + f_F^2 + f_M^2 + 2f_Ff_Mu = 1.$$

In the present analysis, however, the expression on the left (i.e., the correlation between familial environments of siblings) is not constrained to equal one but rather is estimated to be only 0.158 under the general model. This unrealized assumption appears to be part of the reason we obtained considerably larger estimates of f_F and f_M in our previous analysis.

Analysis of all 16 correlations for triglyceride is reported in Table 4. For whatever reason, the parameter t is not significant for triglyceride ($\chi_1^2 = 5.79 - 5.58 = 0.21$, $p > 0.58$), even though its estimate is negative, as expected (-0.021). Therefore, we do not present the analysis with the father–child correlation excluded. Perhaps few fathers in the study were selected primarily for elevated triglyceride levels, thus minimizing the reduction in the father–child correlation. Unlike cholesterol, the sibship common environmental effect is very significant ($\chi_1^2 = 14.72 - 5.58 = 9.14$, $p < 0.003$). The most parsimonious hypothesis with $y = z = 1$, $f_F = f_M$, and $u = t = 0$ (no intergenerational differences, no maternal effects, no assortative mating, and no special father–child correlation) fits very well ($\chi_{10}^2 = 7.23$, $p > 0.68$). In fact, none of the individual components of the parsimonious hypothesis is significant.

The Cincinnati LRC family study

The Cincinnati Lipid Research Clinic (LRC) Princeton School District Family Study (1976–1978), referred to here as the LRC (Morrison et al., 1982), was a part of the National Heart, Lung, and Blood Institute's multicenter collaborative program designed to assess the familial aggregation of lipids and lipoprotein levels (Heiss et al., 1980). Briefly, the Princeton School District Population Study was a two-stage epidemiological survey of lipids, lipoproteins and other coronary heart disease risk factors in a biracial population of school children in grades 1–12 and their parents. Following the first two visits of the prevalence study (Morrison et al., 1978), a subgroup of probands was drawn from this larger prevalence population for the Family Study. All first degree relatives and spouses of selected probands were contacted; sociodemographic data, fasting plasma lipids and lipoproteins, and clinical chemistry measurements were obtained. Probands for the Family Study included both randomly selected subjects and hyperlipidemic subjects (Kelly et al., 1983; Morrison et al., 1982; Tyroler et al., 1979). Here reference is made only to the 160 Caucasian nuclear families ascertained through randomly selected probands. There were a few three-generation families that were split into component nuclear families, avoiding duplications whenver possible. Details of the population studied can be found in Morrison et al (1982) and Laskarzewski et al., (1983).

The multiple regression equation fitted separately to each lipid variable was (Laskarzewski et al., 1983):

$$X = f_2(A,S,C) + g_2(Z) + e_2$$

where $f_2(A,S,C)$ = polynomial in age and sex; also includes the linear effect of oral contraceptive usage (applicable to women), $g_2(Z)$ = linear function of obesity and hematocrit, and e_2 = residual. The lipid variables, adjusted for age, sex, and contraceptive usage (phenotype, P), and the environmental index (I) were defined as (Rao et al., 1982a)

$$P = X - \hat{f}_2(A,S,C)$$

and

$$I = \hat{g}_2(Z).$$

Family data on P and I have recently been analyzed using the same model as given in

TABLE 3. Goodness-of-fit χ^2 values and estimates of parameters under various hypotheses for total cholesterol in the Honolulu Heart Study

Hypothesis	χ^2	d.f.	h^2	c^2	y	z	u	b	f_F	f_M	i	v	t^1
Analysis of all 16 correlations ($t \neq 0$)													
General	7.77	5	0.561	0.032	0.788	1.065	−0.110	0.332	0.045	0.220	1.000	1.000	−0.113
$t = 0$	13.64	6	0.561	0.033	0.794	0.863	−0.110	0.331	0.044	0.221	1.000	1.000	0
$u = 0$	13.18	6	0.561	0.032	0.785	1.065	0	0.334	0.021	0.215	1.000	1.000	−0.114
$b = 0$	11.38	6	0.562	0.033	0.999	1.042	−0.270	0	0.137	0.386	1.000	0.652	−0.108
$f_F = f_M$ and $t = 0$	33.67	7	0.561	0.033	0.762	0.864	−0.110	0.357	0.131	0.131	1.000	1.000	0
$f_F = f_M$ and $t \neq 0$	27.27	6	0.561	0.033	0.762	1.075	−0.110	0.357	0.131	0.131	1.000	1.000	−0.118
$y = z = 1$	8.56	7	0.579	0.029	1	1	−0.133	0.317	0.052	0.243	1.000	0.905	−0.104
Father–child correlation excluded ($t = 0$)													
General	7.77	5	0.561	0.032	0.788	1.065	−0.110	0.332	0.045	0.220	1.000	1.000	0
$u = 0$	13.18	6	0.561	0.032	0.785	1.065	0	0.334	0.021	0.215	1.000	1.000	0
$b = 0$	11.38	6	0.562	0.033	0.999	1.042	−0.270	0	0.137	0.386	1.000	0.652	0
$f_F = f_M$	27.27	6	0.561	0.033	0.762	1.075	−0.110	0.357	0.131	0.131	1.000	1.000	0
$y = z = 1$	8.56	7	0.579	0.029	1	1	−0.133	0.317	0.052	0.243	1.000	0.905	0
(parsimonious)			± 0.047	± 0.009			± 0.082	± 0.071	± 0.045	± 0.067		± 0.208	
$y = z = 1$ and	11.44	8	0.574	0.033	1	1	−0.270	0	0.137	0.386	1.000	0.652	0
$b = 0$			± 0.047	± 0.009			± 0.122		± 0.072	± 0.052		± 0.093	

[1]t = Direct correlational path between the phenotypes of father and child.

TABLE 4. Goodness-of-fit χ^2 values and estimates of parameters under various hypotheses for triglyceride in the Honolulu Heart Study

Hypothesis	χ^2	d.f.	h^2	c^2	y	z	u	b	f_F	f_M	i	v	t^1
General	5.58	5	0.249	0.101	0.913	1.218	0.050	0.334	0.137	0.135	1.000	1.000	−0.021
$t = 0$	5.79	6	0.249	0.101	0.913	1.129	0.050	0.334	0.136	0.136	1.000	1.000	0
$u = 0$	6.75	6	0.249	0.101	0.915	1.218	0	0.331	0.144	0.142	1.000	1.000	−0.021
$b = 0$	14.72	6	0.259	0.100	1.162	1.070	0.032	0	0.222	0.214	1.000	0.730	−0.022
$f_F = f_M$ and $t \neq 0$	5.58	6	0.249	0.101	0.913	1.218	0.050	0.334	0.136	0.136	1.000	1.000	−0.021
$y = z = 1$	6.08	7	0.272	0.099	1	1	0.054	0.325	0.145	0.144	1.000	0.932	−0.007
$y = z = 1$, $f_F = f_M$,	7.23	10	0.267	0.100	1	1	0	0.321	0.152	0.152	1.000	0.928	0
$u = t = 0$ (most parsimonious			± 0.037	± 0.015				± 0.049	± 0.027	± 0.027		± 0.112	

[1]t = Direct correlational path between the phenotypes of father and child.

TABLE 5. Goodness-of-fit χ^2 values and estimates of some parameters for total cholesterol and triglyceride in the Cincinnati LRC (Rao et al., 1982b)

Hypothesis	χ^2	d.f.[1]	h^2	c^2	y	z	u	b	f_F	f_M
Total Cholesterol										
General	4.74	7	0.663	0.078	0.798	0.884	0.183	0.749	0.166	0.252
u = 0	8.08	8	0.663	0.078	0.793	0.885	0	0.736	0.212	0.283
b = 0	13.51	8	0.669	0.079	0.820	0.803	0.229	0	0.520	0.491
y = z = 1, $f_F = f_M$, and u = 0 (most parsimonious)	9.06	11	0.618 ± 0.093	0.070 ± 0.030	1	1	0	0.734 ± 0.184	0.264 ± 0.118	0.264 ± 0.118
Triglyceride										
General	5.92	7	0.230	0.147	1.008	0.759	0.188	0.657	0.138	0.190
u = 0	9.97	8	0.230	0.147	1.008	0.759	0	0.648	0.174	0.216
b = 0	24.24	8	0.252	0.147	1.142	0.322	0.000	0	0.500	0.415
y = z = 1, $f_F = f_M$, and u = 0 (most parsimonious)	10.22	11	0.194 ± 0.092	0.149 ± 0.034	1	1	0	0.648 ± 0.057	0.197 ± 0.058	0.197 ± 0.058

[1]The Cincinnati LRC data yielded estimates of 17 correlations for each variable, 16 from nuclear families and an additional half-sib correlation based on a sample of size 13, adding one extra d.f. under each hypothesis.

Figure 1 (Rao et al., 1982a). Table 5 summarizes the analysis of total cholesterol and triglyceride. For total cholesterol, the general model fits very well ($\chi^2_7 = 4.74$, $p > 0.68$). Unlike the Honolulu Heart Study (see Table 3), assortative mating is not significant ($\chi^2_1 = 8.08 - 4.74 = 3.34$, $p < 0.06$), but the common sibship environmental effect is ($\chi^2_1 = 13.51 - 4.74 = 8.77$, $p < 0.004$). The most parsimonious hypothesis with $y = z = 1$, $f_F = f_M$, and $u = 0$, none of which were individually significant, fits very well ($\chi^2_{11} = 9.06$, $p > 60$).

For triglyceride, the general model fits very well ($\chi^2_7 = 5.92$, $p > 0.54$). Common sibship environmental effect is highly significant ($\chi^2_1 = 24.24 - 5.92 = 18.32$, $p < 0.0001$). Whereas assortative mating is marginally significant ($\chi^2_1 = 9.97 - 5.92 = 4.05$, $p < 0.05$), a parsimonious hypothesis that excludes assortative mating, maternal effects, and intergenerational differences in heritabilities ($u = 0$, $f_F = f_M$, and $y = z = 1$) fits very well ($\chi^2_{11} = 10.22$, $p > 0.48$).

HETEROGENEITY BETWEEN STUDIES

In the present context, between study heterogeneity can be investigated several ways. First, we might systematically compare the correlation estimates from the two studies noting differences whenever they arise. Although such a comparison would suffice to determine the presence or absence of any heterogeneity, it is not likely to indicate what is the underlying source of that heterogeneity. In order to explore between study differences more fully we propose that this preliminary global assessment of heterogeneity be followed by a finer analysis that identifies the source of the correlational heterogeneity in terms of the parameters of the path model. Recently, Rao et al. (1982b) adopted such a strategy to investigate the nature and sources of heterogeneity between these two studies. We summarize those results here.

In testing for correlational differences between the two studies the father–child correlation for CH in the HHS was deleted based upon the analysis presented in Table 3; for total cholesterol this left 15 correlations in the HHS, 3 of which were found to be significantly different between the two studies (see Table 2): between marital indices ($\chi^2_1 = 6.20$, $p < 0.05$), between sibling indices ($\chi^2_1 = 18.22$, $p < 0.001$), and between a child's phenotype and his or her sibling's index ($\chi^2_1 = 5.67$, $p < 0.05$). These differences are consistent with our earlier observation that the significance of the two parameters u and b is reversed in the two studies. For triglyceride only one of the 16 family correlations was significantly different between the HHS and LRC, the correlation between sibling indices ($\chi^2_1 = 22.16$, $p < 0.001$). This difference is consistent with a difference in strength of sibling environment between the two studies and, in fact, although both studies gave significant estimates of b, the estimate of b in the LRC is approximately twice that in the HHS.

For each lipid variable, in order to identify the source of heterogeneity further, we analyzed the two sets of familial correlations simultaneously. As the father–child CH correlation in the HHS had been deleted, this resulted in the analysis of 31 correlations for CH and 32 correlations for TG. Analysis consisted of fitting the multifactorial model given in Figure 1 to the HHS and LRC correlations, although now there are 20 parameters in the model, the 10 parameters of Table 1 subscripted either with a "1" to denote effects for the HHS (e.g., h_1, c_1, etc.) or a "2" to denote effects for the LRC (e.g., h_2, c_2, etc.). By simultaneously treating both sets of correlations we are now able to test specific null hypotheses about the source of between-study heterogeneity through a likelihood ratio test. For example, the difference between the residual χ^2 for the model which constrains $h_1 = h_2$ and the residual χ^2 for the 20-parameter model provides a test for the null hypothesis of equal genetic heritabilities in the two studies. This model and method for testing heterogeneity between multiple studies has been implemented in HPATHMIX, a FORTRAN program written on HARRIS computers. Correlations estimated with PATHMIX (Table 2) were analyzed using HPATHMIX. In HPATHMIX, the χ^2 minimized is the same as discussed before; the only difference is that all individual correlations in both studies now contribute to the χ^2.

Analysis of the 31 correlations on total cholesterol using the 20-parameter model of HPATHMIX is summarized in Table 6. The general model estimating all 20 parameters fits very well ($\chi^2_{11} = 12.24$, $p > 0.33$). Under the null hypothesis of no heterogeneity between the two studies, the 20 parameters were reduced to 10 common ones ($h_1 = h_2$, $c_1 = c_2$, etc.), leading to significant heterogeneity ($\chi^2_{10} = 48.84 - 12.24 = 36.60$, $p < 0.0001$), which is not surprising given the conflicting evidence the two studies provided on b and u (see Tables 3 and 5). We then determined whether b and u were the only source of heterogeneity by allowing the parameters u_1, u_2 and b_1, b_2 to differ, while the remaining parameters were constrained to be equal between the two studies. Consequently a total of 12 parameters were estimated, the fit of this model providing no evidence for any heterogeneity other than for u and b ($\chi^2_8 = 17.73 - 12.24 = 5.49$, $p > 0.69$). What is so impressive is that not even the precision of indices, as measured by i and v, is heterogeneous between the two studies, even though the construction of indices is somewhat different. In terms of the 12-parameter model, intergenerational differences are not significant ($\chi^2_{10} = 18.50 - 12.24 = 6.26$, $p > 0.78$), making the hypothesis of no heterogeneity except for u and b, and no intergenerational differences ($y_1 = y_2 = z_1 = z_2 = 1$) one of the parsimonious hypotheses. When compared to the general model, the parsimonious hypothesis remains tenable even without the effects of assortative mating ($\chi^2_{12} = 29.54 - 12.24 = 17.30$, $p > 0.13$). Estimates of parameters are remarkably similar under the two parsimonious hypotheses, with the only noticeable difference being that the estimate of f_F is halved in the absence of assortative mating.

It is impressive that the only sources of heterogeneity are with respect to two environmental parameters of secondary interest (b and u), which along with f_F, f_M, i, and v are those parameters most likely to be population-specific. The combined evidence of the two studies strongly supports maternal effects. Whereas there is no heterogeneity in f_F or f_M (i.e., $f_{F_1} = f_{F_2}$ and $f_{M_1} = f_{M_2}$: $\chi^2_2 = 1.25$, $p > 0.52$), maternal effects are highly significant under the hypothesis of homogeneity of f_F and f_M ($\chi^2_1 = 19.81$, $p < 0.0001$).

Table 7 summarizes the analysis of the 32 correlations for triglyceride. Again, the general model in 20 parameters fits very well ($\chi^2_{12} = 10.32$, $p > 0.57$), and there is significant heterogeneity between the two studies ($\chi^2_{10} = 43.52 - 10.32 = 33.20$, $p < 0.003$) that is not due to parameters other than b and u ($\chi^2_8 = 15.32 - 10.32 = 5.00$, $p > 0.75$). The last two entries of Table 7 present two parsimonious hypotheses both of which fit very well. The first, which postulates no intergenerational differences in heritabilities ($y_1 = y_2 = z_1 = z_2 = 1$) and permits heterogeneity only in b (i.e., $b_1 = b_2$), fits very well ($\chi^2_{23} = 18.20$, $p > 0.73$; and likelihood- ratio $\chi^2_{11} = 18.20 - 10.32 = 7.88$, $p > 0.71$). The second parsimonious hypothesis that further postulates no assortative mating ($u_1 = u_2 = 0$) and no maternal effects ($f_{F_1} = f_{F_2} = f_{M_1} = f_{M_2}$) also fits very well ($\chi^2_{25} = 21.47$, $p > 0.63$; likelihood-ratio $\chi^2_{13} = 21.47 - 10.32 = 11.15$, $p > 0.57$). The parameter estimates are remarkably similar under the two hypotheses.

In summary, whereas the nontransmitted common sibship environmental effect (b) was the only source of heterogeneity between the two studies for TG, the correlation between marital environments (u) was also indicative of heterogeneity for CH. Differences in the age composition of children in the two studies may partly account for the heterogeneity in b. A nontransmitted sibship environmental effect

TABLE 6. Simultaneous analysis of 31 familial correlations on CH, 15 from HHS, and 16 from LRC: Goodness-of-fit χ^2 values and estimates of some parameters[1]

Hypothesis	χ^2	d.f.	h_1^2	h_2^2	c_1^2	c_2^2	u_1	u_2	b_1	b_2	f_{F1}	f_{F2}	f_{M1}	f_{M2}
General (all 20 parameters estimated)	12.24	11	0.561	0.657	0.033	0.077	−0.110	0.183	0.332	0.744	0.045	0.165	0.220	0.251
No heterogeneity (only the 10 common parameters estimated)	48.84	21	0.592	0.592	0.036	0.036	−0.059	−0.059	0.405	0.405	0.049	0.049	0.223	0.223
No heterogeneity, except that $u_1 \neq u_2$ and $b_1 \neq b_2$	17.73	19	0.588	0.588	0.041	0.041	−0.117	0.205	0.335	0.679	0.058	0.058	0.233	0.233
No heterogeneity, except that $u_1 \neq u_2$, $b_1 \neq b_2$, and $y_1 = y_2 = z_1 = z_2 = 1$	18.50	21	0.594 ± 0.041	0.594 ± 0.041	0.035 ± 0.008	0.035 ± 0.008	−0.143 ± 0.076	0.257 ± 0.147	0.308 ± 0.067	0.643 ± 0.052	0.065 ± 0.040	0.065 ± 0.040	0.250 ± 0.054	0.250 ± 0.054
No heterogeneity, except that $b_1 \neq b_2$, $u_1 = u_2 = 0$, and $y_1 = y_2 = z_1 = z_2 = 1$	29.54	23	0.596 ± 0.041	0.596 ± 0.041	0.033 ± 0.008	0.033 ± 0.008	0	0	0.323 ± 0.060	0.660 ± 0.046	0.036 ± 0.031	0.036 ± 0.031	0.228 ± 0.048	0.228 ± 0.048

[1]The ten parameters defined in Table 1 are subscripted here with a "1" to designate HHS, and a "2" to designate the Cincinnati LRC; for example, $h_1, c_1, \ldots$ correspond to HHS, and $h_2, c_2, \ldots$ correspond to Cincinnati LRC.

TABLE 7. Simultaneous analysis of 32 familial correlations on TG, 16 from each of HHS and the LRC: Goodness-of-fit χ^2 values and estimates of some parameters[1]

Hypothesis	χ^2	d.f.	h_1^2	h_2^2	c_1^2	c_2^2	u_1	u_2	b_1	b_2	f_{F1}	f_{F2}	f_{M1}	f_{F2}
General	10.32	12	0.249	0.252	0.101	0.148	0.050	0.188	0.334	0.657	0.136	0.138	0.136	0.190
No heterogeneity (only the 10 common parameters estimated)	43.52	22	0.253	0.253	0.110	0.110	0.075	0.075	0.391	0.391	0.135	0.135	0.144	0.144
No heterogeneity, except that $u_1 \neq u_2$ and $b_1 \neq b_2$	15.32	20	0.252	0.252	0.110	0.110	0.045	0.211	0.328	0.678	0.138	0.138	0.145	0.145
No heterogeneity, except that $b_1 \neq b_2$ and $y_1 = y_2 = z_1 = z_2 = 1$	18.20	23	0.260 ± 0.034	0.260 ± 0.034	0.107 ± 0.014	0.107 ± 0.014	0.079 ± 0.046	0.079 ± 0.046	0.323 ± 0.046	0.681 ± 0.043	0.139 ± 0.031	0.139 ± 0.031	0.148 ± 0.030	0.148 ± 0.030
No heterogeneity, except that $b_1 \neq b_2$, $y_1 = y_2 = z_1 = z_2 = 1$, $u_1 = u_2 = 0$, and no maternal effects	21.47	25	0.259 ± 0.034	0.259 ± 0.034	0.108 ± 0.014	0.108 ± 0.014	0	0	0.317 ± 0.048	0.678 ± 0.044	0.156 ± 0.024	0.156 ± 0.024	0.156 ± 0.024	0.156 ± 0.024

[1]See footnote to Table 5.

that decreases with child's age would be consistent with the observed difference. This can be examined only when b is made a function of child's age, a possibility that is being investigated at present. Finally, the observed differences in u and b could also be partly due to underestimation of the correlations in the HHS. In any case, the genetic (h^2) and cultural (c^2) heritabilities cannot be considered to be heterogeneous between the populations. Parsimonious hypotheses yield $h^2 = 0.594 \pm 0.041$ and $c^2 = 0.035 \pm 0.008$ for CH, and $h^2 = 0.259 \pm 0.034$ and $c^2 = 0.108 \pm 0.014$ for TG.

BIVARIATE ANALYSIS OF PLASMA LIPIDS

The path models discussed thus far have been primarily concerned with drawing inferences about genetic and environmental influences upon the transmission of a single phenotype. There are several motivations for generalizing path models to allow for the study of the joint transmission of multiple phenotypes. For example, both CH and TG are considered risk factors for artherosclerotic disease. Consequently, understanding the transmission of artherosclerotic disease requires an understanding of the joint transmission of CH and TG, and a determination of whether the phenotypic association between CH and TG represents common genetic effects, common environmental effects or both. Recently McGue et al. (1983) introduced a bivariate path model for the analysis of the joint transmission of two phenotypes. This model has been applied to plasma lipids from the Cincinnati LRC study (McGue, 1983). Here we briefly describe the

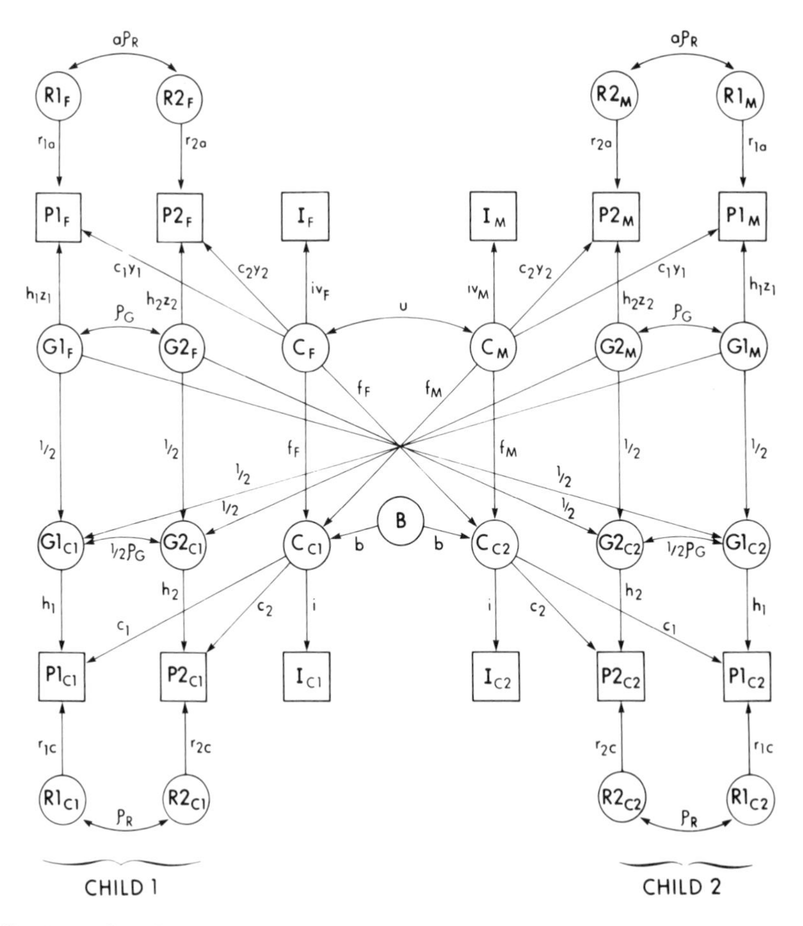

Fig. 2. Bivariate path model showing cultural and biological inheritance for two correlated phenotypes P1 and P2, with corresponding genotypes G1 and G2, a common familial environment C, and residuals R1 and R2. B denotes non-transmitted common sibship environment, and the index I is an estimate of C. Subscripts F, M, C1 and C2 denote father, mother, and two children respectively.

model, review the results of the application of this model to the LRC study, and apply the model to the analysis of the HHS lipid data.

The general bivariate model is represented schematically in Figure 2. The diagram depicts the transmission of two phenotypes, P1 (here taken to be CH) and P2 (here taken to be TG), within a nuclear family. As in the univariate path model, each phenotype is taken to be an additive function of its own genotype (G1 or G2), a single common familial environment (C), and its own residual environment (R1 or R2). As before, we require an estimate of the family environment that here is taken to be the sum of the two indices created separately for CH and TG as previously described (i.e., $I = I_1 + I_2$); this corresponds to taking the first principal component of the two indices. There are 18 parameters in the general model; the four basic parameters given in Table 1 (h, hz, c, cy) are subscripted by "i" to refer to the effect upon the ith phenotype (i = 1,2). The parameters i, u, f_F, f_M, and b are defined as before, while the univariate parameter v has now been generalized to v_F and v_M to allow for differential precision for mothers and fathers, in the estimation of C by I.

The central concern in bivariate path analysis is drawing inferences about the nature of the association between two phenotypes. In the present model, the association between two phenotypes could be the result of common family environmental effects, a correlation (ρ_G) between the two separate genotypes, a correlation (ρ_R in children and $a\rho_R$ in adults) between residual environments, or some combination of these. Specific hypotheses that can be tested using this method include the hypotheses of no genetic association (i.e., $\rho_G = 0$), a single common geneotype (i.e., $\rho_G = 1$), and no residual environmental association (i.e., $a\rho_R = \rho_R = 0$).

For each member of a nuclear family we will observe their cholesterol level (P1) their triglyceride level (P2) and an environmental index (I). These observations will generate 40 familial correlations with unique expectations that can be estimated by repeated execution of PATHMIX, a FORTRAN computer program (Morton et al., 1983). These correlations are then analyzed using BPATHMIX, another FORTRAN computer program that implements the bivariate methodology (McGue et al., 1983). The method of statistical analysis is the same as outlined in Statistical Analysis.

TABLE 8. Observed correlations (r) and associated sample sizes (n) for the two plasma lipids in the Cincinnati LRC and in the HHS

Number	Pair of variables[1]	Cincinnati LRC r	Cincinnati LRC n	HHS Family Study r	HHS Family Study n
1	$P1_F,P1_M$	0.008	80	−0.070	394
2	$P2_F,P2_M$	−0.013	101	−0.008	439
3	$P1_F,P2_F$	0.357	233	0.251*	445
4	$P1_F,P2_M$	−0.107	183	−0.059	836
5	$P1_{C1},P1_{C2}$	0.379	180	0.286	739
6	$P2_{C1},P2_{C2}$	0.266	173	0.140	1070
7	$P1_{C1},P2_{C1}$	0.349	300	0.258	1226
8	$P1_{C1},P2_{C2}$	0.186	267	0.009	1512
9	$P1_F,P1_{C1}$	0.399	135	0.186**	817
10	$P2_F,P1_{C1}$	0.151	88	−0.049	833
11	$P1_F,P2_{C1}$	0.011	73	−0.002	886
12	$P2_F,P2_{C1}$	0.187	103	0.144	986
13	$P1_M,P1_{C1}$	0.246	121	0.304	789
14	$P2_M,P1_{C1}$	0.050	137	0.060	718
15	$P1_M,P2_{C1}$	−0.022	117	0.027	831
16	$P2_M,P2_{C1}$	−0.030	161	0.165	873
17	I_F,I_M	0.164	84	−0.007	402
18	I_F,I_{C1}	0.152	105	0.129	948
19	I_M,I_{C1}	0.265	157	0.203	937
20	I_{C1},I_{C2}	0.492	163	0.153	1003
21	$P1_F,I_F$	0.244	92	0.040	495
22	$P2_F,I_F$	0.385	92	0.195	495
23	$P1_M,I_M$	0.211	131	0.040	445
24	$P2_M,I_M$	0.456	131	0.307	445
25	$P1_F,I_M$	0.083	81	−0.016	445
26	$P2_F,I_M$	−0.030	93	0.001	445
27	$P1_M,I_F$	0.111	81	0.007	445
28	$P2_M,I_F$	0.145	93	0.049	445
29	$P1_F,I_{C1}$	0.133	100	−0.042	930
30	$P2_F,I_{C1}$	0.088	93	0.046	977
31	$P1_M,I_{C1}$	−0.094	130	0.009	846
32	$P2_M,I_{C1}$	−0.024	158	0.044	847
33	$P1_{C1},I_F$	0.028	104	−0.018	799
34	$P2_{C1},I_F$	0.054	97	0.073	912
35	$P1_{C1},I_M$	0.077	126	0.034	712
36	$P2_{C1},I_M$	0.033	141	−0.017	852
37	$P1_{C1},I_{C1}$	0.251	253	0.200	1207
38	$P2_{C1},I_{C1}$	0.402	273	0.270	1282
39	$P1_{C1},I_{C2}$	0.146	255	−0.005	1509
40	$P2_{C1},I_{C2}$	0.217	250	0.054	1842

[1]P1 = CH, and P2 = TG (see Fig. 2).
*The adult correlation between CH and TG is based only on mothers (the corresponding correlation for fathers is −0.032).
**This correlation was excluded in the analysis.

The Cincinnati LRC family study

Table 8 presents the 40 sample correlations and associated sample sizes. Of major interest in the present analysis is the within person correlation between CH and TG, which is estimated to be 0.357 for adults (correlation #3) and 0.349 for children (correlation #7). The bivariate analysis of these correlations is summarized in Table 9. Analysis consisted of first fitting a general model and then several reduced models that allowed tests of specific null hypotheses. The general model is found to fit the observed correlations quite well ($\chi^2_{22} = 17.88$, p = 0.71), with both the genetic and residual environmental correlations estimated to be moderate. We are unable to reject

TABLE 9. Bivariate path analysis of cholesterol (P1) and triglyceride (P2) in the Cincinnati LRC study: Tests of hypotheses and estimates of some parameters

Hypothesis	Residual χ^2	d.f.	Cholesterol h_1^2	Cholesterol c_1^2	Triglyceride h_2^2	Triglyceride c_2^2	u	b	f_F	f_M	ρ_G	ρ_R	a
General	17.88	22	0.690	0.080	0.305	0.185	0.171	0.712	0.200	0.142	0.308	0.251	1.59
$\rho_G = 0$	19.37	23	0.631	0.098	0.261	0.209	0.172	0.751	0.226	0.146	0	0.544	.917
$\rho_G = 1$	19.96	23	0.653	0.088	0.019	0.237	0.172	0.781	0.217	0.140	1	0.215	1.18
a = 1	18.09	23	0.706	0.079	0.316	0.184	0.172	0.712	0.203	0.141	0.236	0.387	1
$\rho_R = 0$	21.30	24	0.658	0.084	0.113	0.214	0.170	0.750	0.194	0.124	0.724	0	1
$\rho_G = \rho_R = 0$	33.15	25	0.504	0.182	0.081	0.324	0.101	0.772	0.305	0.056	0	0	1
Most parsimonious:	23.97	30	0.650	0.069	0.141	0.197	0	0.748	0.182	0.182	0.392	0.272	1
$y_1 = y_2 = z_1 = z_2 = 1$,			$\pm$ 0.087	$\pm$ 0.026	$\pm$ 0.109	$\pm$ 0.052		$\pm$ 0.093	$\pm$ 0.060	$\pm$ 0.060	$\pm$ 0.285	$\pm$ 0.202	
$u = 0$, $f_F = f_M$,													
$v_F = v_M$,													
a = 1													

the hypothesis of no genetic association (i.e., $\rho_G = 0$; $\chi_1^2 = 1.49$, $p = 0.22$), a single common genotype (i.e., $\rho_G = 1$; $\chi_1^2 = 2.08$, $p = 0.15$), no residual environmental correlation (i.e., $\rho_G = 0$; $\chi_2^2 = 3.42$, $p = 0.18$) or no generational differences in the residual environmental correlation (i.e., $a = 1$; $\chi_1^2 = 0.21$, $p = 0.65$). Although either ρ_G or ρ_R could be set equal to zero without significantly increasing the residual χ^2, simultaneously setting both to zero does result in a significant increas in the residual χ^2, ($\chi_3^2 = 15.27$, $p < 0.01$). Some additional form of association between CH and TG other than common family environmental effects is necessary although we are unable to determine statistically whether this additional association is due to a correlation between residual environments, genotypes, or both. It is satisfying to note that the univariate parsimonious hypothesis of Table 5 (no intergenerational differences in heritabilities, no assortative mating, and no maternal effects) both fits the data quite well ($\chi_{30}^2 = 23.97$, $p = 0.77$) and produces parameter estimates that are very similar to those given in the univariate analysis. In the absence of additional information, the best estimates of the genetic and residual environmental correlations are those given under the parsimonious model of $\rho_G = 0.392 \pm 0.285$ and $\rho_R = 0.272 \pm 0.202$.

The Honolulu heart study

Table 8 presents the 40 correlation estimates and sample sizes for the HHS. As is evident from the table, many of the correlations in the HHS are considerably smaller than in the LRC. The estimated within-person correlation between CH and TG is now 0.251 for adults and 0.258 for children. In order to assess the effects of selection upon the analysis we analyzed the correlations three ways: (1) using all 40 correlations, (2) deleting the father–child correlation for CH and the CH–TG correlation for fathers reducing the number of correlations to 39, and (3) deleting all correlations based upon information from the fathers, in which case there are 21 correlations and no information to estimate u, f_F, and v_F. The analysis of these three subsets consisted of fitting the general model and a parsimonious model with no intergenerational differences ($y_1 = y_2 = z_1 = z_2 = a = 1$), no assortative mating ($u = 0$), and no maternal effects ($f_F = f_M$, $v_F = v_M$). The results of these analyses are reported in Table 10. Under the general model there is remarkable consistency across the three subsets for all parameter estimates except for a. For the full data set a was estimated to be 0.370, a

TABLE 10. Bivariate path analysis of cholesterol (P1) and triglyceride (P2) in the Honolulu Heart Study: Tests of hypotheses and estimates of some parameters for several subsets of the 40 observed correlations

Correlations analyzed	Hypothesis[1]	Residual χ^2	d.f.	Cholesterol h_1^2	c_1^2	Triglyceride h_2^2	c_2^2	u	b	f_F	f_M	ρ_G	ρ_R	a
All 40	General	31.35	24	0.556	0.036	0.257	0.072	0.001	0.293	0.171	0.188	0.002	0.397	0.370
	Parsimonious	60.34	31	0.509	0.020	0.270	0.078	0	0.292	0.185	0.185	0.004	0.273	1
				± 0.039	± 0.007	± 0.037	± 0.013		± 0.063	± 0.031	± 0.031	± 0.078	0.066	
Only 39[2]	General	24.97	23	0.560	0.036	0.257	0.072	0.001	0.293	0.172	0.187	0.005	0.394	1.29
	Parsimonious	39.50	30	0.584	0.020	0.270	0.078	0	0.290	0.185	0.185	0.004	0.423	1
				± 0.047	± 0.007	± 0.037	± 0.013		± 0.063	± 0.031	± 0.031	± 0.073	± 0.078	
Only 21[3]	General	7.40	8	0.599	0.038	0.252	0.070	—	0.339	—	0.185	0.099	0.324	1.30
	Parsimonious	14.14	12	0.581	0.027	0.275	0.077	—	0.336	—	0.192	0.090	0.347	1
				± 0.047	± 0.008	± 0.045	± 0.013		± 0.048		± 0.040	± 0.091	± 0.088	

[1]Under the general hypothesis, all relevant parameters are estimsted; see the footnotes below; the parsimonious hypothesis specifies $y_1 = y_2 = z_1 = z_2 = 1$, $u = 0$, $v_F = v_M$, $f_F = f_M$, and $a = 1$.
[2]Father–child correlation for CH is deleted, reducing the total to 39 correlations; also, the adult correlation between CH and TG is based only on mothers (see footnote to Table 8).
[3]All 19 correlations involving fathers are deleted, and therefore, there is no information on u, f_F and v_F; the general hypothesis contains 13 parameters, and the parsimonious one contains 9.

value significantly different from 1 ($\chi_1^2 = 4.65$, $p < 0.05$). In both reduced data sets the estimate of a is slightly, but not significantly, greater than 1 ($a = 1.29$, $\chi_1^2 = 0.46$, $p = 0.50$ and $a = 1.30$, $\chi_1^2 = 0.43$, $p = 0.51$). Apparently the effect of the selection upon fathers is to reduce the residual environmental correlation between CH and TG for fathers. Support for this conclusion comes not only from noting the consistency with which the remaining parameters are estimated over the three subsets, but also by noting that for the Cincinnati LRC a was also estimated to be slightly, but not significantly, greater than one (Table 9).

Comparison of the LRC and the HHS

For the purposes of comparison we will concentrate upon the HHS analysis where father–child CH and father's CH–TG correlations have been deleted. Comparison of the parameter estimates under the parsimonious hypothesis indicate that b, ρ_G, and ρ_R are possible sources of heterogeneity. In the previous section on heterogeneity between studies b was already identified as a source of heterogeneity and we will not discuss it further. For the LRC, both ρ_G and ρ_R are estimated to be moderate and either, although not both, could be set equal to zero without significantly increasing the residual χ^2. For the HHS, ρ_G is estimated to be near, and not significantly different, from zero ($\rho_G = 0.004 \pm 0.073$). In contrast, ρ_R is estimated to be substantial and is significantly different from zero ($\rho_R = 0.423 \pm 0.078$; $\chi_2^2 = 32.41$, $p < 0.001$). This apparent discrepancy between the two studies can be resolved by fixing $\rho_G = 0$ under the parsimonious hypothesis for the LRC data. If $\rho_G = 0$, then the estimate of ρ_R becomes 0.426 ± 0.054 a value clearly consistent with that estimated with the HHS data. Consequently, we would conclude that both studies are consistent with low to zero genetic association and moderate to substantial residual environmental association for the two plasma lipids.

DISCUSSION

Two different family studies of plasma lipids were analyzed by path analysis using the same linear additive model: the Honolulu Heart Study of Japanese–Americans in Hawaii (HHS), and the Cincinnati LRC family study of Caucasians (LRC). Separate analysis of the two studies yielded results that are remarkably similar, with a few minor discrepancies that are consistent with the way in which the two studies were conducted. As the HHS and the LRC contained similar types of information, heterogeneity between the two studies was explored in some depth. Clearly, effect of the nontransmitted common sibship environment (b) was significantly greater in the LRC than in the HHS. But then, a nontransmitted sibship environmental effect that decreases with child's age would give rise to such a difference, as children in the HHS were considerably older and living separately from their parents and sibs. The correlatiron between environments of spouses (u) was also indicative of heterogeneity for total cholesterol. However, whereas u was not found to be significantly different from zero in the LRC, it was found to be significantly less than zero for total cholesterol in the HHS. We are not confident that the negative result is either meaningful or repeatable. By and large, the two studies might be considered homogeneous. In any case, the genetic (h^2) and cultural (c^2) heritabilities cannot be considered to be significantly heterogeneous. Parsimonious hypotheses yield $h^2 = 0.594 \pm 0.041$ and $c^2 = 0.035 \pm 0.008$ for cholesterol, and $h^2 = 0.259 \pm 0.034$ and $c^2 = 0.108 \pm 0.014$ for triglyceride, in good agreement with other investigations such as the total community of Tecumseh, Michigan (Sing and Orr, 1978). It should be pointed out that the finding by Morton et al. (1978) of segregation of a major gene for hypercholesterolemia in the HHS does not invalidate the analysis here based upon a multifactorial model. As this gene has a small frequency, it explains very little of the total phenotypic variance, and consequently practically all of the phenotypic variance remains to be explained by a multifactorial model.

As the two plasma lipids are significantly correlated, a bivariate path analysis was undertaken to test hypotheses about the relationship between the lipids. Specifically, we asked if the correlation between the two lipids was the result of family environment common to both lipids, correlated genotypes, correlated residual environments, or some combination of these factors. The LRC and the HHS studies were anlayzed separately. Both studies were found to be consistent with little genetic association but substantial residual environmental association between the two lipids. About half the within-person association between plasma lipids is due to familial environment with the other half being due to correlated residual environments. This information is clinically valuable in a prevention therapeutic situation, especially when specific environmental risk factors are identified. For example, a particular therapy such as dietary restriction for cholesterol would result in a proportionate reduction in triglyceride, the extent dependent on whether or not the particular therapeutic agent, such as diet, is familial.

It should be remembered that the inferences drawn here are valid only to the degree that the path models and the underlying assumptions, such as linearity, additivity and normality, are valid at least as first order approximations. Cloninger et al. (1983) and Rao et al. (1983) provide ample justification for the continued use of such models and methods. On a related matter, questions have been raised from time to time on the numerical problems associated with simultaneous estimation of many parameters. It should be emphasized that the extent of such problems depends on the particular optimization method used for maximizing a likelihood function, as well as on the number of parameters estimated. Until a few years ago, we were unable to estimate more than ten simultaneous parameters successfully when we used the Newton–Raphson method for maximization. However, the recent development of GEMINI and ALMINI, two numerically efficient general purpose optimization packages (Lalouel, 1979; 1983), constitutes a major improvement in our ability to estimate considerably more parameters. Our computer programs PATHMIX, HPATHMIX, and BPATHMIX are all based on these new packages, and we seldom experience numerical problems.

In studies of disease-related traits such as plasma lipids, a major goal of the clinical biologist is the identification of individual determinants of the phenotype. Mathematical modeling is one of the major tools available for this work. Understanding the extent to which hyperlipidemia is genetically determined and the extent to which it is influenced by other factors, such as diet or physical activity, is a necessary step in the search for better health. We anticipate that path analysis and other models will play an important role in dissecting the determinants of many human characteristics in years to come.

ACKNOWLEDGMENTS

This paper was presented at the Annual Meeting of the American Association of Physical Anthropologists, Eugene, Oregon, April 1982 and was partly supported by NIH and NIMH grants GM 28719, GM17173, and MH31302, and by contract NO-1-HV-2-2914L from the National Heart, Lung and Blood Institute (Lipid Research Clinic's Program), General Clinical Research Center, and the CLINFO center Grant RR-00068-19.

LITERATURE CITED

Cloninger, CR, Rao, DC, Rice, J, Reich, T, and Morton, NE (1983) A defense of path analysis in genetic epidemiology. Am. J. Hum. Genet., in press.

Elston, RC (1975) Correlations between correlations. Biometrika *62:* 133–148.

Fisher, RA (1921) On the probable error of a coefficient of correlation deduced from a small sample. Metron *1:* 1–32.

Garn, SM, Bailey, SM, and Higgens, ITT (1980) Effects of socio-economic status, family line and living together on fatness and obesity. In RM Lauer and RB Shekelle (eds): Childhood Prevention of Atherosclerosis and Hypertension, New York: Raven Press.

Glueck, CJ, Laskarzewski, PM, Rao, DC, and Morrison, JA (1983) Familial aggregation of coronary risk factors. In W Connor and D Bristow (eds): Complications in Coronary Heart Disease. Lippincott, in press.

Gulbrandsen, CL, Morton, NE, Rao, DC, Rhoads, GG, and Kagan, A (1979) Determinants of plasma uric acid. Hum. Genet. *50:* 307–312.

Gulbrandsen, CL, Morton, NE, Rhoads, GG, Kagan, A, and Lew, R (1977) Behavioral, social and physiological determinants of lipoprotein concentrations. Soc. Biol. *24:* 289–293.

Heiss, G, Tamir, I, Davis, CE, Tyroler, HA, Rifkind, BM, Schonfeld, G, Jacobs, D, and Frantz, ID (1980) Lipoprotein-cholesterol distributions in selected North American populations: The Lipid Research Clinics Program Prevalence Study. Circulation *61:* 302–315.

Kelly, KK, Austin, M, Maciolowski, M, Dawson, D, Tyroler, HA, Mowery, R, and Glueck, CJ (1983) The Collaborative Lipid Research Clinics Family Study: Design, ascertainment, lipids and lipoproteins. Hum. Hered., in press.

Krieger, H, Morton, NE, Rao, DC, and Azevedo, E (1980) Familial determinants of blood pressure in Northeastern Brazil. Hum. Genet. *53:* 415–418.

Lalouel, JM (1979) GEMINI—A computer program for optimization of general nonlinear function. University of Utah Technical Report No. 14, December 10, 1979, Salt Lake City, Utah.

Lalouel, JM (1983) ALMINI—A computer program for optimization of nonlinear functions subject to nonlinear constraints. In NE Morton, DC Rao, JM Lalouel (eds): Methods in Genetic Epidemiology. New York: S. Karger, in press.

Lange, K, Westlake, J, and Spence, MA (1976) Extensions to pedigree analysis. III Variance components by the scoring method. Ann. Hum. Genet. *39:* 485–491.

Laskarzewski, PM, Rao, DC, Morrison, JA, Khoury, P, and Glueck, CJ (1983) The Cincinnati Lipid Research Clinic Family Study: Social and physiological determinants of lipids and lipoprotein concentrations. Hum. Hered., in press.

Li, CC (1975) Path Analysis: A Primer. Boxwood Press: Pacific Grove, California.

McGue, M (1983) Bivariate path analysis of plasma lipids. Hum. Hered., in press.

McGue, M, Rao, DC, Reich, T, Laskarzewski, P, Glueck, CJ, and Russell, JM (1983) The Cincinnati Lipid Research Clinic Family Study: Bivariate path analyses of lipoprotein concentrations. Genet. Res., in press.

Morrison, JA, Kelly, KK, Horvitz, R, Khoury, P, Laskarzewski, PM, Mellies, MJ, and Glueck, CJ (1982) Parent–offspring and sibling lipid and lipoprotein associations during and after sharing of household environments: The Princeton School District Family Study. Metabolism *131:* 158–167.

Morrison, JA, Kelly, KK, Mellies, MJ, deGroot, I, and Glueck, CJ (1978) Parent–child associations at upper and lower ranges of plasma cholesterol and triglyceride. Pediatrics *62:* 468–478.

Morton, NE (1974) Analysis of family resemblance. I. Introduction. Am. J. Hum. Genet. *26:* 318–330.

Morton, NE, Gulbrandsen, CL, Rao, DC, Rhoads, GG, and Kagan, A (1980) Determinants of blood pressure in Japanese–American families. Hum. Genet. *53:* 261–266.

Morton, NE, Gulbrandsen, CL, Rhoads, GG, Kagan, A, and Lew, R (1978) Major loci for lipoprotein concentrations. Amer. J. Hum. Genet. *30:* 583–589.

Morton, NE, Rao, DC, Lalouel, JM (1983) Methods in Genetic Epidemiology. New York: S. Karger.

Rao, DC, Morton, NE, and Yee, S (1974) Analysis of family resemblance. II. A linear model for familial correlation. Am. J. Hum. Genet. *26:* 331–359.

Rao, DC, Morton, NE, and Cloninger, CR (1979a) Path analysis under generalized assortative mating. I. Theory. Genet. Res. *33:* 175–188.

Rao, DC, Morton, NE, Gulbrandsen, CL, Rhoads, GG, Kagan, A, and Yee, S (1979b) Cultural and biological determinants of lipoprotein concentrations. Ann. Hum. Genet. *42:* 467–477.

Rao, DC, Laskarzewski, JA, Morrison, P, Khoury, P, Kelly, K, Wette, R, Russell, J, and Glueck, CJ (1982a) The Cincinnati Lipid Research Clinic Family Study: Cultural and biological determinants of lipids and lipoprotein concentrations. Am. J. Hum. Genet., *34:* 888–903.

Rhoads, GG, Gulbrandsen, CL, and Kagan, A (1976) Serum lipoproteins and coronary heart disease in a population study of Hawaii Japanese men. N. Engl. J. Med. *294:* 293–298.

Sing, CF, and Orr, JD (1978) Analysis of genetic and environmental sources of variation in serum cholesterol in Tecumseh, Michigan. IV. Separation of polygene from common environment effects. Am. J. Hum. Genet. *30:* 491–504.

Tyroler, HA, Anderson, T, Chase, G, Ellis, L, Mowery, R, and Valulick, D (1979) The Lipid Research Clinics Population Based Family Study. In C Sing and M Skolnick (eds): Genetic Analysis of Common Diseases: Application to Predictive Factors in Coronary Disease. New York: Alan R. Liss, pp. 647–652.

AMERICAN JOURNAL OF PHYSICAL ANTHROPOLOGY 62:51–59 (1983)

Genetic Analysis of Manic-Depressive Illness

D.H. O'ROURKE, P. McGUFFIN, AND T. REICH
Department of Anthropology, University of Utah, Salt Lake City, Utah 84112 (D.H.O.); Institute of Psychiatry, Maudsley Hospital, London (P.M.); Department of Psychiatry (P.M., T.R.), and Department of Genetics, Washington University School of Medicine and the Jewish Hospital of St. Louis (T.R.), St. Louis, Missouri 63110

KEY WORDS Bipolar affective disorder, Segregation analysis, Combined model

ABSTRACT Two threshold models, a single locus model, and two combined models are fitted to data on familial incidence of bipolar affective disorder in 194 nuclear families ascertained through a bipolar proband. The relative fit of alternative tramsmission models is tested by a likelihood ratio chi-square with the degrees of freedom defined by the difference in the number of parameters estimated by each model. All parameters are estimated by the method of maximum likelihood. The simplest threshold model, permitting only a single background familial correlatiron, is found to provide a statistically poorer fit than any of the alternative models, and may be rejected as a model for the etiology of bipolar affective disorder. The four remaining models are statistically indistinguishable. It is suggested, however, that the involvement of a major locus in the etiology of this disorder deserves further scrutiny since any of the models incorporating a major locus, with or without a multifactorial background, are consistently associated with greater likelihoods than the complex threshold model.

It is also noted that diagnostic criteria are critical in the analysis. In the present study, relatives of probands are considered affected if a diagnosis of bipolar or unipolar affective disorder is present. When only bipolar relatives are considered affected, none of the transmission models may be rejected. Finally, the results of these analyses are found to be independent of the ascertainment parameter.

The interaction of biology and behavior is central to the investigations of biological anthropology. Of particular interest are those behavioral phenotypes that relate to fitness and have a demonstrable biological basis. Manic-depressive illness is just such a phenotype. So called "bipolar affective disorder" (BP) is characterized by extreme mood swings between elation (mania) and depression, whereas unipolar (UP) disorder is characterized by depressive episodes alone. Mania comprises (1) elated, unstable, and fluctuating moods; (2) pressure of speech, occasionally associated with rhyming or punning and circumlocution; (3) flight of ideas, grandiosity, and increased distractability; (4) increased motor activity and energy output; (5) exercise of poor judgement; and (6) increased aggressive or sexual behaviors. In contrast, a depressive episode is marked by (1) depressed mood; (2) psychomotor retardation; (3) deterioration of thought processes, which may include paranoid ideation, accusatory hallucinations, and delusions; (4) suicidal ideation; (5) decreased sexual interests; and (6) abnormalities of physiological function such as loss of appetite, weight loss, and sleep disturbance (Balis et al., 1978).

Although females exhibit UP disorder nearly twice as often as males, recent surveys indicate that the population prevalence of BP disorder is roughly comparable between the sexes (see Reich et al., 1983 and McGuffin and Reich, 1983 for reviews). Population prevalence (K_P) estimates for BP disorder range from a low of 0.0036 to a high of 0.022 (Slater and Roth, 1969).

Family, twin, and adoption studies have demonstrated the importance of genetic factors

Received June 5, 1982; accepted March 16, 1983.

in the etiology of bipolar affective disorder. However, the mode of inheritance of this non-Mendelian trait has remained obscure. One advance for genetic studies of affective disorders was the diagnostic distinction proposed by Leonhard (1959) between bipolar and unipolar affective disorders. Subsequent studies by Angst (1966) and Perris (1966) illustrated that family members of UP probands showed increased frequencies of UP disorder but no increase of BP disorder. In contrast, relatives of BP probands showed elevated rates of affectation for BP disorder and, in most studies (Angst 1966; Gershon, et al. 1975a,b; Gershon and Leibowitz, 1975; Helzer and Winokur 1974; Smeraldi, et al. 1977; Winokur and Clayton 1967) increased frequencies of primary depression without episodes of mania, suggesting that depressed relatives of BP probands should be considered affected whether or not a manic episode has been observed. This dichotomy and diagnostic procedure is supported by data suggesting that conversion to BP disorder after several depressive episodes is reasonably high (perhaps approaching 18%, e.g., Akiskal et al. 1977). Consequently, in the present study, relatives of BP probands diagnosed as suffering from primary depression are considered affected.

MATERIALS AND METHODS

The data come from three family studies of BP disorder conducted by members of the Psychiatry Department, Washington University School of Medicine, St. Louis (Winokur et al., 1969, Helzer and Winokur, 1974, and unpublished data on 103 black families ascertained through a BP proband) between 1965 and 1975. In all cases the diagnostic criteria used were either those of Feighner et al. (1972), or closely comparable criteria developed in the Psychiatry Department, which were the forerunners of the Feighner criteria. These three studies were first analyzed separately and remarkably similar results obtained. Therefore, in subsequent analyses, the results of which are presented here, the studies of the St. Louis group have been pooled to include data on 194 BP probands, their sibs, and parents.

Various forms of three genetic transmission models have been fitted to these data: two multifactorial threshold models, a single major locus model, and two combined models; that is, a major locus with multifactorial background contributing to liability to the disease.

The models

Multifactorial (MF). Discrete, presence/absence traits, such as BP disorder may be the result of polygenic or multifactorial inheritance where specific combinations of alleles at multiple loci, together with environmental contributions, additively determine liability to the disease (Falconer, 1965, 1967; Crittenden, 1961; Reich et al., 1972, 1975). Liability to the disease is considered normally distributed (or transformable to a normal distribution) within the population (Fig. 1), and only those individuals above the liability threshold are affected. It may be emphasized that even where evidence exists for unequal effects (e.g., a few loci of major effect), the assumption of equal effects of all factors underlying the liability distribution is reasonable since it leads to accurate estimates of risks to the disease as well as response to selection (Falconer, 1960, 1965). Moreover, the assumption of normality of the liability distribution is independent of the equality of effects of underlying factors since it depends only on the distribution of phenotypic values. This assumption, then, merely reflects the choice in scale of measurement. The Central Limit Theorem assures rapid approach to normality as long as many underlying factors are relevant (Cloninger et al., 1983; Reich et al., 1972). Under this mode of inheritance, relatives of affected probands will show a distribution on the liability scale with a higher mean than the general population and, hence, a greater proportion of affected individuals.

Since the distribution of liability is considered normal, we may take it as a standardized distribution with the unit of measurement the standard deviation, the threshold at zero, and

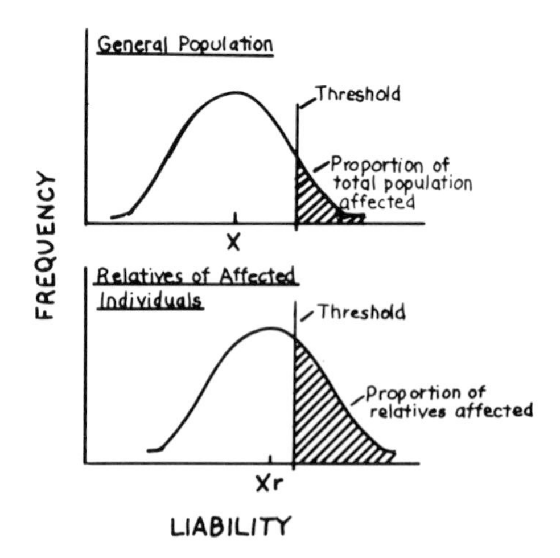

Fig. 1. Multifactorial model with single threshold. Probands are ascertained from pool of affected in general population. Relatives of probands show an increase in mean of the distribution on the liability scale as well as an increase in the proportion of affected individuals.

the total variance in liability for the population equal to one. Taking x_P and x_R as the normal deviates of the thresholds for probands and relatives of probands for their respective distributions on the liability scale, the correlation between relatives on the liability scale is (Reich et al., 1972):

$$r = \frac{x_P - x_R \sqrt{[1 - (x_P^2 - x_R^2)(1 - (x_P/a))]}}{a + x_R^2 (a - x_P)}$$

where a is the deviation of the proband mean from the population mean. This correlation and the threshold values (or the prevalences in the general population and relatives of probands) determine the MF model under the following additional assumptions (Falconer, 1965; Reich et al., 1972): (1) all genetic and environmental causes of the disease may be combined into a single continuous variable termed the liability; (2) the population is divided into at least two recognizable classes by one or more thresholds; (3) genes that are relevant to the etiology of the disease are separately of small effect relative to the total variation; (4) although not required (see above), genes are assumed to act additively in order to allow data on sibs to be combined with those from parents and offspring; (5) there are multiple environmental contributions to etiology that act additively; and (6) for purposes of genetic parameter estimation, common environmental effects among relatives are negligible.

If the assumptions of the model are met, r is independent of the population mean and is invariant over changes in definition of the threshold. Further, this model "applies only to those diseases whose genetic component is multifactorial, or if there are few genes, where these have effects that are small in relation to the non-genetic variation" (Falconer, 1965:53). In those cases where the genetic component is a single locus of major effect, an alternative model must be employed.

Single major locus (SML). Such traits may be the result of a single diallelic locus (A and a) where each of the three genotypes may be incompletely penetrant (Fig. 2). This model is defined by the a gene frequency ($q = 1 - p$) and the penetrance values (f_1, f_2, f_3) for each of the three genotypes (AA, Aa, and aa, respectively), where the penetrances are the probability of each genotype exhibiting the phenotype under study. Collectively, the penetrance values are referred to as the penetrance vector.

James (1971) showed that morbid risk data permit estimation of only three parameters; the population prevalence (K_P), the additive genetic variance (V_A), and the dominance var-

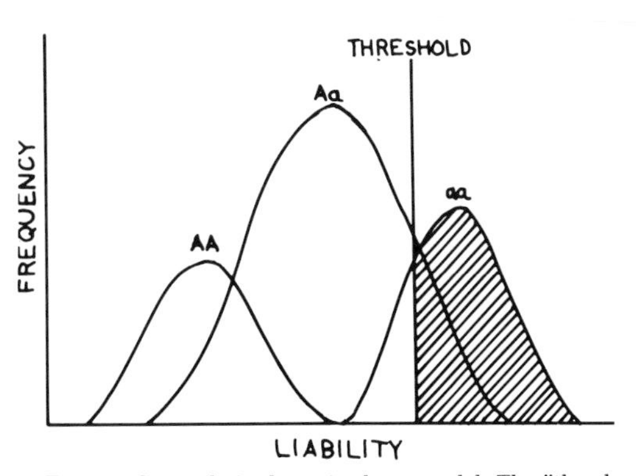

Fig. 2. General single major locus model. The "threshold," analagous to that in multifactorial model, is defined by the penetrance vector; the probability of each genotype exhibitng the phenotype under study.

iance (V_D) irrespective of the number of classes of relatives of probands studied. Thus, a parameter problem is encountered. Determination of these three values is insufficient to estimate the four underlying parameters uniquely (Kempthorne, 1957):

$$K_P = p^2f_1 + 2pqf_2 + q^2f_3$$

$$V_A = 2pq\,[q(f_3 - f_2) + p(f_2 - f_1)]^2$$

$$V_D = p^2q^2[f_1 - 2f_2 + f_3]^2.$$

As noted by James (1971) incidence of a disease in relatives of a proband is a function of K_P and covariance (COV_R) between relatives,

$$K_R = K_P + \frac{COV_R}{K_P},$$

where the covariance between relatives is the sum of weighted proportions of the two genetic variances (i.e., $COV_R = uV_A + vV_D$). The weights (u and v) are simply the probabilities that relatives share one particular allele and both alleles at a locus identical by descent, respectively. For the class of relatives considered here these weights are 0.5 and 0 for parents and offspring and 0.5 and 0.25 for sibs.

Fortunately, the parameter problem just discussed may be resolved by maximizing the information contained in nuclear family units as opposed to population prevalences and correlations between relatives. Fishman et al. (1978) have shown that examination of the joint prevalence probability for sibs (e.g., both sibs in a sibship of size two are affected) in conjunction with the observed parental phenotypes permits unique estimation of the underlying parameters of the single major locus model. Here, the phenotype is dichotomized as affectational status where an affected individual may be coded as 1 and an unaffected individual as 0. Fishman et al. (1978) prove the theorem that the joint probability of affectation in sibs, conditional on parental phenotypes for a presence–absence trait, e.g.,

$$P_s\,(1,1|i,j) = V_A/2 + V_D/4 + K_P^2 + (\gamma_i + \gamma_j)K_PV_A + \gamma_i\gamma_j\,K_PV_A\sigma_D + (\gamma_i + \gamma_j)\,J_1/4 + \gamma_i\gamma_j\,J_2$$

with

$$J_1 = (p - q)\,\alpha\,(V_A + V_D) + 3V_A\sigma_D + V_D\sigma_D,$$

$$J_2 = [V_A + V_D/2 + (p - q)\alpha\sigma_D/2]^2 - V_A/2$$

and

$$\alpha = p(f_2 - f_1) + q(f_3 - f_2)$$

$$\gamma_0 = -(1 - K_P)^{-1}$$

$$\gamma_1 = K_P^{-1}$$

uniquely specifies the four underlying parameters if K_P, $P_o\,(1|1)$ and one each of $P_o\,(1|i,j)$ and $P_s\,(1,1|i,j)$ are known with $V_A \neq 0$. The conditional probabilities $P_o\,(1|1)$ and $P_o\,(1|i,j)$ refer to the prevalence of affected offspring of affected individuals and prevalence of affected offspring given parental mating type, respectively.

In fact, it is further shown that $P_s\,(1,1|i,j)$ need only be known for one mating type as is usually the case (especially for rare recessive traits). Thus, the present study utilizes complex segregation analysis to examine the pattern of distribution of the trait within nuclear families and to maximize the $\log_e$ likelihood over families, rather than just correlations between pairs of family members.

Combined model (CM). Finally, the underlying genetic mechanism may represent a combination of the two preceding models; a major locus and an additive multifactorial background that augments correlations between relatives and contributes to liability (Morton and MacLean, 1974).

Under this model, the presence of a dichotomous trait (x) is the result of an effect due to a major locus (g), background correlations between relatives (c) due to multiple additive genetic factors, and an environmental component.

$$x = g + c + e.$$

The environmental effect may be subdivided into common familial and random environments, each with a normal distribution of mean zero and separate, estimated variances such that their sum is the variance of e. Similarly, c, the polygenic effect may be subdivided into the midparent breeding value as well as the individual deviation from this value with each normally distributed around mean zero.

The major locus is defined by its mean value across genotypes, the degree of dominance, frequency, and its displacement (that is, the effect of substituting one allele for the other, see Morton and MacLean, 1974 for complete discussion of the model). It should be noted that since the means of c and e are taken to be zero, the phenotypic mean is the same as the mean of the major locus (Fig. 3). This is intuitively obvious since everyone in the population has one of the three genotypes defined by the diallelic major locus.

It has been suggested that such a model may provide the final resolution to the mode of inheritance of BP disorder (Reich et al., 1983; McGuffin and Reich, 1983). The present study is the first to assess the fit of a combined model to the familial distribution of manic-depressive illness.

Table 1 summarizes the models used and parameters estimated in the present study. All parameters are estimated by the method of maximum likelihood. The two multifactorial models differ only in estimation of correlations between relatives. The first model (MF1) assumes a single familial background correlation, whereas MF2 estimates six possible correlations. By definition the familial correlations are assumed to be zero for the single major locus model, although the three genotypic penetrances are allowed to vary between the sexes. For the two combined models the six specific

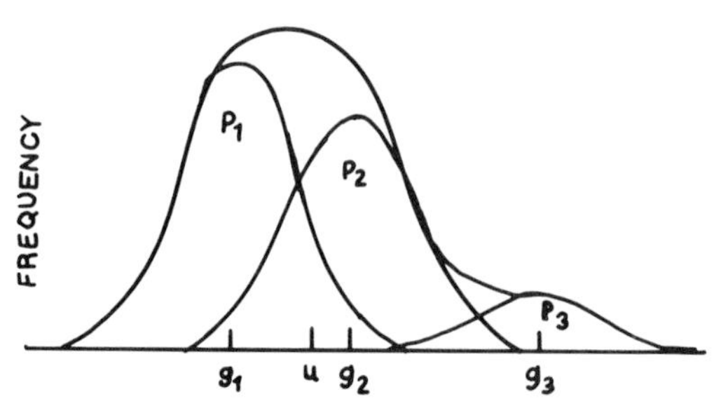

Fig. 3. The combined model. Phenotype is determined by a major locus in conjunction with multifactorial background. See text for discussion.

TABLE 1. Models used and parameters estimated

Model	Parameters estimated	Number of parameters
Multifactorial (MF1)	♂ threshold, ♀ threshold, 1 familial correlation	3
Multifactorial (MF2)	♂ threshold, ♀ threshold, 6 familial correlations[1]	8
Single major locus (SML)	gene frequency (q), 6 penetrances[2]	7
Combined (CM1)	gene frequency (q), 6 penetrances,[2] 1 familial correlation	8
Combined (CM2)	gene frequency (q), 6 penetrances,[2] 6 familial correlations[1]	13

[1]Parent/offspring = male–male, male–female, female–female; Sib/sib = male–male, male–female, female–female.
[2]Males = f_{1m}, f_{2m}, f_{3m}; females = f_{1f}, f_{2f}, f_{3f}.

penetrance values are allowed to vary but only CM2 estimates multiple familial correlations. In all these preliminary analyses an ascertainment probability (π) of 0.0001 is assumed (i.e., the probability of an individual being affected *and* a proband is small ($\pi \rightarrow 0$), so that we have effectively single ascertainment with one proband per family).

One advantage of the present approach is that a single likelihood value defines the fit of each model used. It is thus possible to test statistically the relative fit of nested models to a single data set. The appropriate test follows a chi-square distribution.

$$\chi^2 = -2 (L_1 - L_2),$$

where L_1 and L_2 are the observed $\log_e$ likelihoods for the two models being compared, and the degrees of freedom are defined by the difference in number of parameters estimated by each model. The possibility exists, then, to select the most adequate model to describe the familial distribution of BP disorder, and test the significance of its superiority over competing models.

RESULTS AND DISCUSSION

Table 2 summarizes the results of the fitting of the two multifactorial models to the data. It is worth noting that while both MF models predict population prevalences of approximately 0.03 and 0.04 for males and females, respectively, the observed K_P for the St. Louis data was 0.034 for both sexes. The small difference between the estimated values, and their symmetry around the observed prevalence, testifies to the roughly equal prevalence rates in both sexes for this disorder. The moderately high familial correlations estimated by both models is not surprising; it reflects the well known elevation in morbid risk to BP disorder

TABLE 2. Parameter estimates under multifactorial models

Estimates	MF1	MF2
Correlations[1]	0.43	0.21 0.57 0.58 0.33 0.30 0.37
K_p (M/F)	0.028/0.042	0.03/0.04
Likelihood	−700.1	−687.2
X^2_5	25.8	($p < 0.01$)

[1]Parent/offspring = MM, MF, FF; sib/sib = MM, MF, FF.

in first degree relatives of probands. Somewhat surprising are the high mother–offspring correlations in MF2, suggesting some form of maternal effect. Indeed, such a pattern led earlier workers to postulate X-linkage for BP disorder (e.g., Winokur and Tanna, 1969; Mendlewicz and Fleiss, 1974; Mendlewicz et al., 1972). However, X-linkage for BP disorder has not been unequivocally demonstrated in general, and several cases of father–son transmission are now known. Thus, undue importance should not be ascribed to the high mother–offspring correlations reported here. Perhaps of greater interest is the fact that the MF2 model, with the full complement of familial correlations, provides a significantly better fit than MF1.

A summary of the results from the single major locus model are presented in Table 3. With an estimated gene frequency of 0.031, the male and female population prevalences of 0.029 and 0.038 are virtually identical to those estimated under the assumptions of multifactorial inheritance (see Table 2). Given the relatively low gene frequency, all of the homozygotes are predicted to be affected. However, considerably larger proportions of females than males are affected as heterozygotes, while a slight excess of male sporadics is predicted under this model.

TABLE 3. Parameter estimates under single major locus model

Estimates	Males		Females
Penetrances			
f_1	0.010		0.006
f_2	0.306		0.530
f_3	0.999		0.999
K_p	0.029		0.038
q		0.031	
Likelihood		−683.9	

TABLE 4. Percentage contribution to pool of affected by genotype and sex for SML

Genotype	Male	Female
"Sporadics" (p^2f_1/K_p)	32.8	14.7
Heterozygotes ($2pqf_2/K_p$)	63.8	82.8
Homozygotes (q^2f_3/K_p)	3.3	2.5

In Table 4 these values, taken together with the predicted K_P values, reveal more clearly the subtle sex differences. The majority of affected individuals of both sexes are predicted to be heterozygous under this model, although nearly one-third are found to be sporadics among males.

Finally, two separate combined models were fitted to these data. Table 5 summarizes these results. A somewhat surprising result is obtained. For both CM1 and CM2 the familial correlation estimates reduce to essentially zero, approximating a simple, single major locus model. In fact, all remaining parameter estimates are found to be exactly the same as those noted earlier for the SML. Given the lack of evidence for appreciable background familial correlations contributing to liability in the combined models, the proportions affected by sex and genotype under these models are the same as those seen earlier for the SML (see Table 4).

Having reviewed the analyses of these data sets through the implementation of multifactorial and complex segregation models, it is now possible to test statistically the relative efficacy of each model. Table 6 presents the results of these tests. MF1 provides a statistically poorer fit than either of the combined models. Since CM1 is, in effect, a single major locus model, the latter would be viewed as providing a statistically better fit than MF1 as well. This is not a trivial point since the SML is not, technically, a model nested within the parameterization of the MF models and, hence, a direct statistical comparison is precluded. Here the clear cut distinctions end. ALthough it is not possible to compare MF2 and CM1 directly (they estimate equal numbers of parameters and, hence, have no degrees of freedom), MF2 is not significantly different from CM2. Since the two combined models and the SML model are statistically indistinguishable, all four of the remaining models must be considered to fit the data equally well. It is not surprising, then, that numerous authors, using transmis-

TABLE 5. Parameter estimates under combined models

Estimates	CM1	CM2
Correlations[1]	0.0	0.0 0.0 0.0 0.0 0.0 0.0
Penetrances[2]	0.010 0.304 0.999 0.006 0.530 0.999	0.010 0.304 0.999 0.006 0.530 0.999
K_p (M/F)	0.029/0.038	0.029/0.038
q	0.031	0.031
Likelihood	−683.9	−683.9

[1]Parent/offspring = MM, MF, FF; sib/sib = MM, MF, FF.
[2]Male = f_{1m}, f_{2m}, f_{3m}; female = f_{1f}, f_{2f}, f_{3f}.

TABLE 6. Direct comparison of likelihood values for models fitted to bipolar affective disorder family studies

Models and likelihoods[1]	Chi-square	df	Probability
MF1 (700.1)—CM1 (638.9)	32.4	5	<0.01
MF1 (700.1)—CM2 (683.9)	32.4	10	<0.01
MF2 (687.2)—CM2 (683.9)	6.6	5	>0.05

[1]Absolute values of likelihoods in parentheses.

sion models similar to those described here, have argued for single major locus inheritance (Winokur et al., 1969; Crowe and Smouse, 1977; Gershon, 1975a,b,c), including X-linkage (Mendlewicz and Fleiss, 1974; Mendlewicz et al., 1972; Winokur and Tanna, 1969; Baron, 1977), as well as multifactorial transmission (Slater and Tsuang, 1968; Bucher and Elston, 1981; Bucher et al., 1981; Baron, 1980). The present analyses suggest that a combined model may be added to the list, bearing in mind that the combined models employed provided essentially a single major locus with little evidence for the importance of a multifactorial background.

Although only MF1 may be statistically rejected using these data, it is worth noting that the presence of a major locus, with or without a polygenic background, has the highest likelihood based on the combined St. Louis data. This at least suggests that it is a reasonable working hypothesis. Indeed, such a working hypothesis is not inconsistent with recent reports of linkage of a disease susceptibility locus for BP disorder to the major histocompatability complex on the short arm of chromosome 6, although these reports need to be verified (e.g., Smeraldi and Bellodi, 1981; Weitkamp et al., 1981). For example it has been shown recently (Suarez and Van Eerdewegh, 1981) that misspecification of the mode of inheritance of a disease susceptibility locus may give spuriously high lod scores for linkage to a known marker locus. Since it is by no means clear whether the genetic etiology of BP disorder is multifactorial, the result of a single major locus, or some combination, much less whether the putative single locus is dominantly or recessively transmitted, such reports must await further confirmation and testing.

The results presented here are concordant with some previous studies but at variance with others. Gershon et al. (1975b, 1976; Gershon and Leibowitz, 1975) fitted MF and SML models to incidence data on UP and BP disorders in the Jewish population of Jerusalem. In these studies, both MF and SML models were found to predict incidence of affectation adequately in relatives of probands. Although these results are similar to those presented here, methodological differences preclude a more rigorous comparison of the studies.

Gershon et al. (1975b, 1976; Gershon and Leibowitz, 1975) were in effect testing the validity of the Leonhard (1959) dichotomy of polarity for affective disorders. Thus, they used a two-threshold model for both the MF and SML analyses with BP and UP disorders representing narrow and broad forms of a disease on a single liability scale. Given this major distinction between the underlying assumptions and diagnostic criteria used by Gershon and colleagues and the present study, it is interesting that the results and inferences are similar. It should be noted, however, that in the present case the less complex form of the MF model may be statistically rejected. Moreover, the greater likelihood associated with those models containing a major locus suggests that further examination of the role of a major locus in this disorder is warranted.

In contradistinction to this position, Bucher and Elston (1981) and Bucher et al. (1981) have explicitly rejected the notion of involvement of a major locus in the etiology of BP disorder. Once again, methodological differences make direct comparisons of analyses and inferences difficult. The methodology used by Bucher and Elston (1981) and Bucher et al. (1981) is Elston's approach to segregation analysis (e.g., Elston 1980; Elston and Stewart 1971). This is a different parameterization of the SML than that used in the present study, and may account for some of the differences in result. Additionally, rather than testing the relative fit of competing models directly, Bucher and colleagues test a Mendelian model against a general unrestricted model. That is, one in which the transmission probabilities of each phenotype are not constrained by Mendelian values.

By not being able to reject this general unrestricted model, Bucher and coworkers suggest that a major locus hypothesis for BP disorder is untenable. However, no model other than the SML was tested. Moreover, the nature of the test suggests that deviations of the individual data sets from Hardy–Weinberg equilibrium conditions could give rise to the results obtained.

Further work in this area is imperative since Bucher and coworkers used some of the same data utilized in the present report (Helzer and Winokur, 1974; Winokur et al., 1969), but came to dramatically different conclusions. In addition, the effects of small sample size, and the pooling of separate studies that utilize different diagnostic criteria on the results of genetic model fitting to incidence data require further investigation (see O'Rourke et al., 1982 for brief review).

The approximation of both combined models to a major locus model in the present analysis not only suggests the importance of a major gene in this behavioral disorder, but raises once again the question of whether this putative disease susceptibility locus is autosomal or X-linked. Evaluation of this question is outside the scope of the present paper. However, Van Eerdewegh et al. (1980) have recently examined the fit of three separate X-linked threshold models to familial data on bipolar affective disorder. Two of these models were single threshold models and differed only in regard to whether individuals suffering unipolar depression were considered affected or unaffected. All families were ascertained through a BP proband. Using three separate data sets, Van Eerdewegh et al. (1980) found that X-chromosome single locus models did not consistently describe the distribution of affected relatives of BP probands. They note that their results may suggest etiological heterogeneity. We are currently reexamining the St. Louis data in light of the present reports' suggestion of major locus involvement in order to evaluate the relative efficacy of autosomal versus X-chromosome transmission models. The result of this research will appear elsewhere.

Finally, bipolar affective disorder is known to have a variable age of onset. Unfortunately, adjustment for age-dependent penetrance is not possible with the analytical techniques used in this analysis. We are, however, modifying the routines employed here to permit adjustment for the age of onset distribution in order to evaluate what effect this may have on the reported results. The age distribution of the probands and family members is such that we do not believe age adjustment will radically alter the results.

SUMMARY AND CONCLUSIONS

While these preliminary analyses suggest that a major locus may be important in the etiology of bipolar affective disorder, several cautionary notes need be appended.

(1) We have not tested the "fit" of a specific model to a set of data. Rather, we have attempted to evaluate the *relative* value of several transmission models to account for the distribution of bipolar affective disorder in families.

(2) Although we have carried out these analyses assuming an ascertainment probability (π) of 0.0001, we cannot know that this is, in fact, the correct value. To test our assumption, we have reanalyzed the St. Louis data under a range of different values of π (range = 0.0001–0.999), and the results are relatively invariant.

(3) Although BP disorder has a variable age of onset, we were unable to use these genetic models on age-corrected data. However, since computational constraints placed a limit on sibship size to five, no family member under the minimum age at risk (15 years) was included in the analysis. Moreover, in those sibships where more than five members had entered the risk period, the youngest were deleted so as to maximize subjects who had passed through the maximum period of risk. Since no birth order effect has been noted for BP disorder, we do not believe this introduces any bias.

(4) Diagnostic criteria are clearly important. In the present study, we have considered relatives affected if they were diagnosed as BP or primary depression only. When only BP relatives are considered affected, none of the five models used are statistically distinguishable. Given the possibility of genetic heterogeneity in BP disorder, further work in this area is imperative.

(5) These analyses were carried out using computer programs developed at Washington University School of Medicine in St. Louis under the supervision of the third author. These programs are but one set out of several that allow analysis of data under the assumptions of multifactorial inheritance or single major locus involvement through complex segregation analysis. Different programs use different algorithms and different parameterizations of the models. It is not yet clear that they all produce concordant results. Until some degree of standardization is achieved, results may be considered tenative.

(6) Finally, results and hypotheses generated by analysis of data using one analytic technique should be continually retested using different analytical methods based on different sets of assumptions. We are currently analyzing these data using several different methodologies in order to more fully elucidate the genetic mechanisms involved in bipolar affective disorder.

ACKNOWLEDGMENTS

We gratefully acknowledge the assistance of Dr. John Helzer, and Mr. Joe Mullaney for his able programming skills. This work was supported in part by USPHS grants MH31302, MH14677, MH25430, and GM28067, and an MRC (U.K.) Fellowship (Dr. P. McGuffin).

LITERATURE CITED

Akiskal, HS, Djenderedjan, AH, Rosenthal, RH, and Khani, MK (1977) Validating criteria for inclusion in the bipolar affective group. Am. J. Psychiat. *134:*1227–1233.

Angst, J (1966) Zur Atiologie and Nosologie endogener depressiver Psychosen. In: Monographien aus der Neurologie und Psychiatrie, No. 122. Berlin: Springer Verlag, pp. 1–118.

Balis, GU, Wurmer, L. McDaniel, E (1978) Clinical Psychopathology. Boston: Butterworth.

Baron, M (1977) Linkage between an X-chromosome marker (deutan color blindness) and bipolar affective illness. Arch. Gen. Psychiat. *34:*721–725.

Baron, M (1980) Genetic models of affective disorder: application to twin data. Acta Genet. Med. Gemellol. *29:*289–294.

Bucher, KD, and Elston, RC (1981) The transmission of manic-depressive illness. I. Theory, description of the model and summary of results. J. Psychiat. Res. *16:*53–63.

Bucher, KD, Elston, RC, Green, R, Whybrow, P, Helzer, J, Reich, T, Clayton, P, and Winokur, G (1981) The transmission of manic-depressive illness. II. Segregation analysis of three sets of family data. J. Psychiat. Res. *16:*65–78.

Clonginger, CR, Reich, T, Suarez, BK, Rice, J, Gottesman, II (1983) The principles of psychiatric genetics. In M. Shepherd (ed): Handbook of Psychiatry, vol. 5: The Scientific Foundation of Psychiatry. Cambridge: Cambridge University Press (in press).

Crittenden, LB (1961) An interpretation of familial aggregation based on multiple genetic and environmental factors. Ann. N.Y. Acad. Sci *91:*769–780.

Crowe, RR, and Smouse, PE (1977) The genetic implications of age-dependent penetrance in manic-depressive illness. J. Psychiat. Res. *13:*273–285.

Elston, RC (1980) Segregation analysis. In JH Mielke and MH Crawford (eds): Current Developments in Anthropological Genetics. New York: Plenum, pp. 327–354.

Elston, RC, and Stewart, J (1971) A general model for the genetic analysis of pedigree data. Hum. Hered. *21*:523–542.
Falconer, DS (1960) Introduction to Quantitative Genetics. London: Oliver and Boyd.
Falconer, DS (1965) The inheritance of liability to certain diseases estimated from incidence among relatives. Ann. Hum. Genet. *29*:51–76.
Falconer, DS (1967) The inheritance of liability to diseases with variable age of onset with particular reference to diabetes mellitus. Ann. Hum. Genet. *31*:1–20.
Feighner, JP, Robins, E, Guze, SB, Woodruff, RA, Winokur, G, and Munoz, R (1972) Diagnostic criteria for use in psychiatric research. Arch. Gen. Psychiat. *26*:57–63.
Fishman, PM, Reich, T, Suarez, BK, and James, JW (1978) A note on the essential parameters of the two-allele autosomal locus model. Am. J. Hum. Genet. *30*:283–292.
Gershon, E, Baron, M, and Leckman, J (1975a) Genetic models of the transmission of affective disorders. J. Psychiat. Res. *12*:301–317.
Gershon, E, and Leibowitz, T (1975) Sociocultural and demographic correlates of affective disorders in Jerusalem. J. Psychiat. Res. *12*:37–50.
Gershon, E, Mark, A, Cohen, N, Belizon, N, Baron, M, and Knobe, K (1975b) Transmitted factors in the morbid risk of affective disorders: a controlled study. J. Psychiat. Res. *12*:283–299.
Gershon, E, Bunney, WE, Leckman, JF, Van Eerdewegh, M, and DeBauche BA (1976) The inheritance of affective disorders: a review of data and hypotheses. Behav. Genet. *6*:227–261.
Helzer, J, and Winokur, G (1974) A family interview study of male manic-depressives. Arch. Gen. Psychiat. *31*:73–77.
James, JW (1971) Frequency in relatives for an all-or-none trait. Ann. Hum. Genet. *35*:47–49.
Kempthorne, O (1975) An Introduction to Genetic Statistics. New York: John Wiley and Sons, Inc.
Leonhard, K (1959) Aufteilung der endogenen Psychosen. Akademie Verlag Berlin.
McGuffin, P, and Reich, T (1983) Psychopathology and Genetics. In: HE Adams and PB Sutker (eds): Comprehensive Handbook of Psychopathology. New York: Plenum Press (in press).
Mendlewicz, J, and Fleiss, J (1974) Linkage studies with X-chromosome markers in bipolar (manic-depressive) and unipolar (depressive) illness. Biol. Psychiat. *9*:261–294.
Mendlewicz, J, Fleiss, J, and Fieve, RR (1972) Evidence for X-linkage in the transmission of manic-depressive illness. J. A. M. A. *222*:1624–1627.
Morton, NE, and MacLean, CJ (1974) Analysis of family resemblance. III. Complex segregation analysis of quantitative traits. Amer. J. Hum Genet. *26*:489–503.
O'Rourke, DH, Gottesman, II, Suarez, BK, Rice, J, and Reich, T (1982) Refutation of the general single-locus model for the etiology of schizophrenia. Am. J. Hum. Genet. *34*:630–649.
Perris, C (1966) A study of bipolar (manic-depressive) and unipolar recurrent depressive psychoses. Acta Psychiat. Scand. Suppl. No. 194.
Reich, T, James, JW, and Morris, CA (1972) The use of multiple thresholds in determining the mode of transmission of semicontinuous traits. Ann. Hum. Gen. *36*:163–184.
Reich, T, Cloninger, CR, and Guze, SB (1975) The multifactorial model of disease transmission: I. Description of the model and its use in psychiatry. Br. J. Psychiat. *127*:1–10.
Reich, T, Cloninger, CR, Suarez, B, and Rice, J (1983) Genetics of the affective psychoses. In: JK Wing (ed): Handbook of Psychiatry. London: Cambridge University Press, in press.
Slater, E, and Roth, M (1969) Clinical Psychiatry. New York: Williams and Wilkins.
Slater, E, and Tsuang, MT (1968) Abnormality on paternal and maternal sides: observations in schizophrenia and manic depression. J. Med. Genet. *6*:197–199.
Smeraldi, E, and Bellodi, L (1981) Possible linkage between primary affective disorder susceptibility locus and HLA haplotypes. Am. J. Psychiat. *138*:1232–1234.
Smeraldi, E, Negri, F, and Melic, AM (1977) A genetic study of affective disorders. Acta Psychiat. Scand. *56*:382–398.
Suarez, BK, and Van Eerdewegh, P (1981) Type I (insulin dependent) diabetes mellitus. Is there strong evidence for a non-HLA linked gene? Diabetologia *20*:524–529.
Van Eerdewegh, MM, Gershon, ES, and Van Eerdewegh, PM (1980) X-chromosome threshold models of bipolar manic-depressive illness. J. Psychiat. Res. *15*:215–238.
Weitkamp, LR, Stancer, HC, Persad, E, Flood, C, and Guttormsen, S (1981) Depressive disorders and HLA: a gene on chromosome 6 thay can affect behavior. N. Engl. J. Med. *305*:1301–1306.
Winokur, G, and Clayton, P (1967) Family history studies. I. Two types of affective disorders separated according to genetic and clinical factors. Recent Adv. Biol. Psychiat. *9*:35–50.
Winokur, G, Clayton, P, and Reich, T (1969) Manic-Depressive Illness. St. Louis: Mosby.
Winokur, G, and Tanna, J (1969) Possible role of X-linked dominant factors in manic-depressive disease. Dis. Nerv. Sys. *30*:89.

AMERICAN JOURNAL OF PHYSICAL ANTHROPOLOGY 62:61–66 (1983)

Social Selection and Evolution of Human Diseases

SHOZO YOKOYAMA
Departments of Psychiatry and Genetics, Washington University, St. Louis, Missouri 63110

KEY WORDS Social Selection, Human Diseases, Evolution

ABSTRACT Disease incidences in human populations depend on etiology of the disease, the fitness of individuals, and demographic changes of the population. The fitness of an individual is determined not only by the disease but also by other factors such as cultural and social reaction to the disorder and demographic changes of the population. Social selection studies the effect of the social behavior on the incidence of a trait.

In studies of Huntington's disease, it has been shown that the fitness of the normal sibling of an affected individual is reduced as much as that of the affected individual himself or herself. A similar social effect has been observed for mental retardation. Thus, even if an individual has a normal genotype, mate finding and fertility may be changed considerably by the presence of affected family members. At the present time, the way in which genetic variabilities are maintained is poorly understood even for clearcut genetic diseases. Studies of social selection indicate that such information should be acquired by considering both the nature of the disease and its social effect.

Disease incidence in a population depend on the etiology of the disease, the fitness of individuals, and demographic structure of the population. In traditional population genetics, the fitness of an individual has been quantified by using only his or her genotype or phenotype. For some diseases, however, the fitness of an individual is determined not only by the disease but by other factors such as cultural and social reactions to the disorder (Reed and Neel, 1959; Lindelius, 1970; Buck et al., 1975; Yokoyama et al., 1980a,b). Therefore, the traditional population genetics approaches are not sufficient to understand evolutionary changes of human diseases at the population level.

Huntington's disease, which is caused by a dominant gene, is probably the best example of such fitness interactions. Reed and Neel (1959) estimated the relative fitnesses of affected to nonaffected sibs to be 1.01 and 1, respectively. This and other studies (Reed and Palm, 1951; Shokeir, 1975) show that this disease is either neutral or slightly favored in heterozygous condition. However, when compared to the general population, the fitness of the heterozygote was 0.81 and, therefore, the fitness of normal sib is about 0.8 (Reed and Neel, 1959).

This example clearly shows that even if an individual has a normal genotype or phenotype, the fitness is considerably affected by the existence of the affected individual in the family. This type of fitness interaction has been termed social selection (Wallace, 1976; Templeton, 1979; Yokoyama, 1980a,b, 1981a, Yokoyama and Templeton, 1980).

In this review, I shall describe an example of social selection for mental retardation. The roles of social selection on the mechanism for the maintenance of genetic diseases in human populations will also be discussed.

FITNESS INTERACTION WITH MENTAL RETARDATES

Reed and Reed (1965) have obtained one of the most extensive data bases of mental retardation and have reported pedigrees over several generations. To detect a social reaction, we compared the marriage frequencies of normal individuals with or without mental retardates in their nuclear family (Yokoyama et al., 1980). Marriage frequency is simply defined as the frequency of individuals who ever married.

Received June 5, 1982; accepted March 16, 1983.

The marriage frequency is only one component of the fitness determination process, but it still gives us an important information on the fitness of an individual.

We used only the data from the original study, that is, the data collected during the years 1911–1918, since marriage status is often unclear at the bottom of the pedigrees. Furthermore, by restricting our attention to the original data, trends due to changing social factors will be minimized. Mentally retarded individuals are classified as L (slight), M (moderate), and S (severe), corresponding to IQ scores of 50–69, 20–49, and less than 20, respectively, and individuals with IQ score > 69 are classified as N. Not all individuals are given IQ scores in the pedigrees, and we assume that the class "unknown mental ability" is N (Reed and Reed consider this the most likely), and that "mentally deficient" individuals without IQ scores are of class L. An individual who is "unknown as to whether or not there are children" is excluded from analysis, and the resulting individuals are organized into nuclear families. The resulting number of nuclear families of the parental types N × N, N × L, and L × L were 7,397, 758, and 132, respectively.

We distinguish normal individuals with no sibs or with normal sibs, with at least one L sib, with at least one M sib, with at least one L sib and one M sib, and with at least one S sib. These classes are denoted by NN, LL, MM, LM, and SS, respectively. Thus, normal individuals are classified into different groups depending upon their sex and upon parental and sibs' phenotypes.

The marriage frequencies for the different combinations of sex and parental and sib phenotypes are shown in Table 1 which is modified from Table 1 in Yokoyama et al. (1980). The mating behavior of the two sexes differ. For females, the marriage frequency is weakly affected by sibs' phenotypes, but decreases by about 10% when both parents are retarded. For males, the magnitude of reduction in the marriage frequency becomes larger as the severity of retardation increases. For example, the marriage frequency without any retarded sib is about 80%, as compared to 65% when at least one S sib is present. As in the case of females, there is a direct effect due to parental phenotype. Thus, in males, the marriage frequency is affected by both parents and sibs. The logistic regression analysis shows that the effects of sex, parental phenotype, and sibling's phenotype are all significant (for details, see Yokoyama et al., 1980).

TABLE 1. Marriage frequencies for the combination of sex of an individual and different parental and siblings' phenotypes

Sex of normal individual	Parental phenotypes	Sibs' phenotypes NN	LL	LM	MM	SS	Total
Female	N × N	0.839	0.850	0.781	0.820	0.794	0.838
	N × L	0.826	0.821	0.816	0.556	0.895	0.814
	L × L	0.769	0.688	0.800	1.000	1.000	0.737
	Total	0.838	0.834	0.795	0.786	0.823	0.834
Male	N × N	0.798	0.752	0.960	0.788	0.673	0.791
	N × L	0.763	0.642	0.806	0.643	0.500	0.693
	L × L	0.857	0.711	0.556	—	1.00	0.714
	Total	0.797	0.728	0.863	0.761	0.648	0.781

It should be noted that Reed and Reed (1965) compared the reproduction of 208 normal persons from sibships with a retardate (IQ less than 70) to that of 208 normal first cousins from a sibship without a retardate. They have shown that the reproductive fitness is the same for the two groups (2.4 children per person). Thus, there seems to exist a discrepancy between the present results and those of Reed and Reed. The difference may be due to the definition of mental retardation. If the mental retardation is defined as IQ less than 70, then the marriage frequencies of a normal person with at least one retarded sib are 4,931/6,030 (81.8%) and 1,765/2,257 (78.2%), respectively. Although a homogeneity test is significant ($\chi^2[1] = 13.52$, $p < 0.001$), the difference between these marriage frequencies is small. Thus, people's response to the disease depends upon its severity and it is important to distinguish the retardates according to their severity in order to assess the social response of mental retardation.

It may seem possible to explain the decrease in the marriage frequency of a normal male with at least one S sib by considering that a normal individual with severely retarded sibs is more likely to have a lower IQ than a normal individual with less severely retarded sibs. Because of this lower IQ (even though higher than 70), the normal individual with a severely retarded sib has a lower marriage frequency than those with less severely retarded sibs. However, this explanation does not seem likely. Severe mental retardation is more likely to be caused by a chromosomal aberration or by a single gene, whereas mild retardation seems to be caused by many loci (Morton et al., 1977).

Accordingly, the correlation of IQ scores between normals and their mildly retarded sibs would be expected to be higher than that between normals and severely retarded sibs. At any rate, our analysis suggests that the difference in the marriage frequencies of normal individuals with and without retarded family members is likely due to social responses to the trait.

MODELS OF SOCIAL SELECTION

When a genetically affected individual modifies the reproductive fitness of other family members, the disease incidence in a whole population will be different from that without such a fitness interaction. Mathematical analyses are helpful in evaluating the effects of such social reactions on disease incidences. In the following, I shall consider two different situations: (1) affected parents modify the fitness of offsprings, and (2) affected person modifies the fitnesses of the parents or those of sibs. For the latter case, several reproductive compensation models will be discussed. The data analysis on mental retardation shows that the sociocultural reactions are affected by sex, and by parental and siblings' phenotypes. In the mathematical modelings of the social selection, however, I do not consider these effects simultaneously because of the algebraic complexity.

Fitness modification by affected parents

Assume that a mutant gene *a* and its wild-type allele *A* are segregating with respective frequencies p and q in the adult population. Let us denote the frequencies of *aa*, *aA*, and *AA* in adults by u, v, and w, respectively. Assume that *aa* always develop a disease and that *AA* is normal. Assume also that h is the degree of dominance of the trait so that, with probability h, *Aa* will also be affected by the disease. There are three parental combinations: both parents are normal, one parent is affected, and both are affected. An individual can therefore be classified into nine groups, that is, three genotypes and three parental combinations, before an individual develops the trait.

Let γ be the magnitude of fitness loss of an individual due to the trait. Furthermore, let $\beta/2$ and β be the magnitude of fitness loss when an individual has one and two affected parents, respectively. We assume that the fitness of an individual is determined multiplicatively by the individual's and parental phenotypes.

After the development of the disease and selection, the frequencies of the three genotypes in the next generation is given by

$$\begin{aligned} u' &= p^2(1-\gamma)(1-H_1)/W \\ v' &= 2pq(1-h\gamma)(1-H_2)/W \\ w' &= q^2(1-H_3)/W \end{aligned} \tag{1}$$

where $H_1 = \beta[u/p + \{h(1-\gamma)v\}/\{2(1-h)p\}]$, $H_3 = \beta[\{h(1-\gamma)v\}/\{2(1-h\gamma)q\}]$, $H_2 = (H_1 + H_3)/2$, and $W = p^2(1-\gamma)(1-H_1) + 2pq(1-h\gamma)(1-H_2) + q^2(1-H_3)$ [Eq.(1) in Yokoyama, 1981a].

For this model, a sufficient condition for both alleles to be maintained in the population is

$$-2\gamma/(1-2\gamma) < \beta < -2\gamma/(1-\gamma) \tag{2}$$

Under this condition, rather high equilibrium gene frequency is achieved (Yokoyama and Rice, in press).

When $\beta > -2\gamma/(1-\gamma)$, the frequency of allele *a* is low and the counterbalance due to mutation becomes important. Suppose that the irreversible mutation from *A* to *a* occurs with rate α per generation. When the forces of selection and mutation are balanced, the equilibrium frequency of alleles *a*, $\hat{p}$, is given by

$$\hat{p} \simeq \begin{cases} \alpha/hs & \text{for } 0 < h \leq 1 \\ \sqrt{\alpha/s} & \text{for } h = 0 \end{cases} \tag{3}$$

where $s = \gamma + \beta(1-\gamma)/2$ (Yokoyama, 1981a).

Equation 3 shows that the equilibrium gene frequency $\hat{p}$ decreases as the magnitude of so cial response to a disease increases.

Reproductive compensation and the resulting fitness interaction

Reproductive compensation was first postulated by Fisher (in Race, 1944) in its relationship to Rh polymorphism. Since then, the hypothesis has been studied by several population geneticists (Li, 1953; Lewontin, 1953; Feldman, et al., 1969; Hagy and Kidwell, 1972; Templeton, 1979). In these studies, reproductive compensation was modeled due to a direct response to the deaths caused by an autosomal genetic disease. The effect of reproductive compensation has also been studied considering X-linked recessive diseases (Templeton and Yokoyama, 1980).

The phenomena of reproductive compensation clearly belong to social selection because parents of affected offspring modify their reproductive strategy. Reproductive compensation is not a sole social selection for the fitness modification of the parents by affected offspring. Selective abortion related to an affected offspring provides another example. Thus, genetic counseling that is aimed at preventing suffering and disease also creates a form of social selection.

As already noted, reproductive compensation has been proposed to explain the maintenance of a stable Rh polymorphism. Two phenotypes, Rh(+) and Rh(−), in the Rh blood group are inherited as a dominant trait: genotypes DD and Dd are Rh(+) and dd is Rh(−). A heterozygous individual Dd is at risk with respect to hemolytic disease when his or her mother is dd, Rh(−). This incompatibility selection between a mother and a child effectively operates against heterozygotes and it is expected that the system creates an unstable equilibrium. However, the observed frequencies of d genes in European and American populations are about 40% (e.g., see Glass, 1949). If the Rh polymorphism is stable, heterozygote advantage in some form may be needed (Cavalli-Sforza and Bodmer, 1971). In the following, I shall give an alternative explanation which would account for population with high frequency of d alleles (Yokoyama, 1981b).

Let x and 1 − x be the respective frequencies of d and D alleles in adults. Suppose that each family produces exactly the same number of live offspring. Without such a reproductive compensation, the matings dd(♀) x DD(♂) and dd(♀) x Dd(♂) would produce smaller number of offspring due to the incompatibility selection. Thus, the present model contains the fitness modification of parent by affected offspring (see Glass, 1950). Furthermore, let s be the probability of Dd individuals dying due to hemolytic disease from a dd mother. Then, the disadvantage of Dd individuals from the mating dd(♀) x DD(♂) disappears. The selective disadvantages of Dd and dd individuals from the mating dd(♀) x Dd(♂) can be determined as follows. The relative viabilities of Dd and dd offspring are $1 - s$ and 1, respectively. Because of the exact number of offspring in each family, the relative fitnesses of Dd and dd in the mating are given by $(1 - s)/(2 - s)$ and $1/(2 - s)$, respectively. This is a special case of the models of Levin (1967) and Feldman et al. (1969), where they studied stability condition of the equilibrium allele frequencies and did not pay any attention to the transient state.

The change in the frequencies of d genes in successive generations is given by

$$\Delta x = sx^3(1 - x)/[4(2 - s)] \tag{4}$$

Thus, under this model, the frequency of d genes always increases. This is an alternative possibility to the mutation-drift hypothesis proposed by Haldane (1942), where he considered that the high frequency of d genes in an American population is probably unstable and that the d gene is in the process of being eliminated.

To study the effect of fitness modification of parents due to affected offspring, we have also studied reproductive compensation for a X-linked recessive lethal disease (Templeton and Yokoyama, 1980). Let X denote a normal X chromosome, and X′, an X chromosome bearing an allele that is lethal when homozygous or hemizygous. The adult population consists of three genotypes: normal females (XX), carrier females (XX′), and normal males (XY). Let Q is the frequency of XX′ females. Suppose the population practices family planning such that the average family size is two for all families irrespective of female genotype. If the female is a carrier, the sex ratio in her living offspring is two females to one male.

Assuming the equal mutation rates in both sexes,

$$Q' = (2/3)Q/(1 + Q/3) + 2u \tag{5}$$

where the prime indicates the next generation and u is the rate of mutation to the lethal gene per generation (see Eq. 5 in Templeton and Yokoyama, 1980). Using Eq. (5), the equilibrium frequency of carrier females, $\bar{Q}$, is given by 6u, but it is 4u without compensation (Haldane, 1935). Thus, the reproductive compensation increases the incidence of females carriers by about 1.5 times.

Now suppose that the cultural and social environment of the population is such that male offspring are desired. For simplicity, we consider that all couples continue to produce children until they have one male offspring and then stop. In this case, the average numbers of offspring for XX and XX′, mothers are given by 2 and 3, respectively (see Table 1 in Templeton and Yokoyama, 1980). Then, the equilibrium frequency of carrier females is given approximately by $\sqrt{2u}$ and the frequency of carrier females increases considerably [see Eq. (14) in Templeton and Yokoyama, 1980].

The impact on carrier female frequency is far greater when the compensation stems from the desire to have male offspring than when it is simple compensation to replace a lost offspring. Templeton and Yokoyama (1980) assumed that the fixed proportion of females continue producing children until they have at least one normal son. In practice, however, the mode of cultural inheritance may not be strict. It is most likely that some may not plan their families with the object of having at least one son even though they were raised in a family with this bias, or vice versa.

It is then possible to construct a two-state Markov chain model. Let α be the probability that the female offspring of compensating parents will not compensate. Furthermore, let β be the probability that the female offspring of noncompensating parents will compensate. Under this model, when $\alpha = 0$ the equilibrium frequency of heterozygous females is given by $\sqrt{2u}$, whereas when $\alpha \neq 0$ it is given by $2[\{\beta + (2\alpha + \beta)(\alpha + \beta)\}/\{\alpha(\alpha + \beta)\}]u$ (Yokoyama and Templeton, 1982). The latter quantity reduces to 4u only when $\beta = 0$, as expected. Furthermore, the equilibrium frequency of heterozygous females depends not only on the intensity of the reproductive compensation, but also on the time of mutational change. It has been shown that the frequency ranges from 4u without compensation to $\sqrt{2u}$ or $\sqrt{3u}$ with strict compensation. The frequency $\sqrt{2u}$ is achieved when mutation occurs in mature germ cells, whereas $\sqrt{3u}$ is achieved when mutation occurs in early development of germ cells (Yokoyama, 1982). Thus, the frequency of carrier females depends strongly on the sociocultural environment.

All of these informations of the equilibrium frequency of carrier females have important implications on detection of newly arisen mutations. Using new biochemical techniques, it is now possible to distinguish newly arisen mutation from old mutant. It is thus possible to test Haldane's equilibrium theory (1935) for X-linked lethals: one-third of all affected males should be fresh mutants if the mutation rates in both sexes are the same.

Zatz et al. (1976), Zatz et al. (1977), and Danieli et al. (1977) have supported Haldane's prediction in their studies of Duchenne muscular dystrophy. However, Biggs and Rizza (1976) reported the value 0.023 for the frequency of fresh mutation among affected males for hemophilia A. Similarly, Franke et al. (1976) observed the value 0.085 for Lesch-Nyhan disease. These values are much lower than the expected value of 1/3.

The fraction of affected males due to new mutation, G, is given by $u/(\overline{Q}/2 + u)$. When there is no reproductive compensation $\overline{Q} = 4u$ and $G = 1/3$ as noted by Haldane (1935). When $\overline{Q} = \sqrt{2u}$, $G = 2u/(\sqrt{2u} + 2u)$ which is 0.0045 for $u = 10^{-5}$ and .0014 for $u = 10^{-6}$. When $\overline{Q} = \sqrt{3u}$, it is 0.0036 and 0.0012 for $u = 10^{-5}$ and $u = 10^{-6}$, respectively. Thus, the proportion of affected males due to fresh mutation ranges from the order of 10^{-3} to 1/3 depending upon the intensity of reproductive compensation.

DISCUSSION

Data analyses of Huntington's disease and mental retardation show that the change in population incidences of a genetic disease cannot be studied by simply comparing the reproductive fitnesses of different phenotypes. To obviate that difficulty, the concept of social selection has been proposed. Traditional reproductive compensation models clearly belong to social selection. Differences and similarities between the present social selection and other concepts such as kin selection, gene-culture coevolution, and cultural evolution have been discussed elsewhere (Yokoyama, in press).

In this study, two types of fitness interactions have been reviewed. It is also possible to construct more general models where individuals are distinguished depending upon whether they have at least one affected nuclear family member or not. Such situations have been studied considering dominant, recessive, and X-linked diseases, separately (Yokoyama, in press). Let us denote the amount of reduction in mating success or in fertility for an individual who has at least one affected nuclear family member by w. Suppose that the frequency of deleterious genes is q and that the irreversible mutation rate from the wild type allele to the deleterious gene is u per generation.

For a dominant disease, let h be the amount of an intrinsic fitness reduction due to the disease. Then, the equilibrium value of q, $\hat{q}$, is given by $u/[h + (1 - h)w]$. For example, consider the case of Huntington's disease, where the estimated values of h and w are 0.01 and 0.2, respectively (Yokoyama and Templeton, 1980). Thus, the disease would increase if it were not for social selection which is the only directional force keeping the allele rare in the population. For a recessive disease, the equilibrium gene frequency under selection and mutation is given by $\hat{q} = \sqrt{u/(1 + w)}$. Thus, the gene frequency increases as the value of w decreases. For an X-linked recessive lethal trait, the equilibrium frequency of the lethal genes in females is given $2u/(1 + w)$ (Yokoyama, in press).

All of the three examples show that the gene frequency is affected strongly not only by the intrinsic deleterious effect of a disease, but also by the behavioral response to the presence of a family member with a disease. It is important to note that the effect of social selection is intensified for a larger progeny size and a smaller population size (Yokoyama, 1980b; in press).

Both Huntington's disease and mental retardation reduce the reproductive fitness of an individual due to an affected nuclear family member. For some traits, a positive effect may occur. Autosomal recessive albinism among Amerindians in the southwestern United States and in Panama may be such an example. The incidence of albinism in these populations range from 1 in 140 to 1 in 3,750 and these high frequencies may be due to such a positive social selection (see Yokoyama and Morgan, submitted).

Information on reproductive fitness come from a variety of disciplines such as anthropology, population biology, psychology, and sociology, which seldom communicate with each other. Such collaborative studies seem to be essential for the studies of disease incidences and for assessing mutational damage in human populations.

ACKNOWLEDGMENTS

This work was supported by grants GM-28672 and MH-31302 from the United States Public Health Service with Washington University.

LITERATURE CITED

Biggs, R, and Rizza, CR (1976) The sporadic case of haemophilia A. Lancet *2*:431–433.

Buck, C, Hobbs, GE, and Simpson, H (1975) Fertility of the sibs of schizophrenic patients. Br. J. Psychiatr. *127*:235–239.

Cavalli-Sforza, LL, and Bodmer, WF (1971) The Genetics of Human Populations. San Francisco: WH Freeman.

Danieli, GA, Mostacciuolo, ML, Bontante, A, and Angelini, C (1977) Duchenne muscular dystrophy: A population study. Hum. Genet. *35*:225–231.

Feldman, MW, Nabholz, M, and Bodmer, WF (1969) Evolution of the Rh polymorphism: A model for the interaction of incompatibility, reproductive compensation, and heterozygote advantage. Am. J. Hum. Genet. *21*:171–193.

Franke, U, Felsenstein, J, Gartler, SM, Migeon, BR, Dancis, J, Seegmiller, JE, Bakay, F, and Nyhan, WL (1976) The occurrence of new mutants in the X-linked recessive Lesch-Nyhan disease. Am. J. Hum. Genet. *28*:123–137.

Glass, B (1949) The relation of Rh incompatibility to abortion. Am. J. Obstet. Gynecol. *57*:323–332.

Glass, B (1950) The action of selection on the principal Rh alleles. Am. J. Hum. Genet. *2*:269–278.

Hagy, GW, and Kidwell, JF (1972) Effect of amniocentesis, selective abortion, and reproductive compensation on the incidence of autosomal recessive disease. J. Hered. *63*:185–188.

Haldane, JBS (1935) The rate of spontaneous mutation of a human gene. J. Genet. *31*:317–326.

Haldane, JBS (1942) Selection against heterozygosis in man. Ann. Eugen. *11*:333–340.

Levin, BR (1967) The effect of reproductive compensation on the long-term maintenance of the Rh polymorphism: The Rh crossroad revisited. Am. J. Hum. Genet. *19*:288–302.

Lewontin, RC (1953) The effect of compensation on populations subject to natural selection. Am. Naturalist *87*:375–381.

Li, CC (1953) Is Rh facing a crossroad. Am. Naturalist *87*:257–261.

Lindelius, R (1970) A study of schizophrenia. Acta Psychiatr. Scand. (Suppl). *216*:1–126.

Morton, NE, Rao, DC, Lang-Brown, H, MacLean, J, Bart, RD, and Lew, R (1977) Colchester revisited: A genetic study of mental defect. J. Med. Genet. *14*:1–9.

Race, RR (1944) Some recent observations on the inheritance of blood groups. Br. Med. Bull. *2*:165.

Reed, SC, and Palm, JD (1951) Social fitness versus reproductive fitness. Science *113*:294–296.

Reed, EW, and Reed, SC (1965) Mental retardation: Family studies. Philadelphia and London: WB Saunders.

Reed, TE, and Neel, JV (1959) Huntington's chorea in Michigan. 2. Selection and mutation. Am. J. Hum. Genet. *11*:107–136.

Shokeir, MHK (1975) Investigation of Huntington's disease in the Canadian prairies. II. Fecundity and fitness. Clin. Genet. *7*:349–353.

Templeton, A (1979) A frequency dependent model of brood selection. Am. Naturalist *114*:515–524.

Templeton, AR, and Yokoyama, S (1980) Effect of reproductive compensation and the desire to have male offspring on the incidence of a sex-linked lethal disease. Am. J. Hum. Genet. *32*:575–581.

Wallace, DC (1976) The social effect of Huntington's chorea on reproductive effectiveness. Ann. Hum. Genet. *39*:375–379.

Yokoyama, S (1980a) The effect of social selection on population on dynamics of rare deleterious genes. Heredity *45*:271–280.

Yokoyama, S (1980b) The effect of variable progeny size and social selection on population dynamics of rare lethal genes. Soc. Biol. *27*:70–77.

Yokoyama, S (1981a) Social selection in human populations. I. Modification of the fitness of offspring by an affected parent. Am. J. Hum. Genet. *33*:407–417.

Yokoyama, S (1981b) Family size and evolution of Rh polymorphism. J. Theor. Biol. *92*:119–125.

Yokoyama, S (1982) Cultural inheritance of having male offspring and the incidence of a sex-linked lethal disease. Soc. Biol. *28*:315–327.

Yokoyama, S Theories of social selection in human populations. Am. J. Hum. Genet. (in press).

Yokoyama, S, and Morgan, K The maintenance of the polymorphism of tyrosinase-positive albinism in American indian populations by social selection. (Submitted.)

Yokoyama, S, and Rice, JP Social selection in human populations. II. Deterministic analyses on the modification of the fitness of offspring by affected parents. Soc. Biol. (in press).

Yokoyama, S, and Templeton, A (1980) The effect of social selection on the population dynamics of Huntington's disease. Ann. Hum. Genet. *43*:413–417.

Yokoyama, S, and Templeton, A (1982) Effect of cultural inheritance of reproductive compensation on the incidence of a sex-linked lethal disease. J. Theor. Biol. *99*:389–395.

Yokoyama, S, Rice, JP, and Yokoyama, RW (1980) The effect of social selection due to familial mental retardation on the marriage frequency of normal individuals. Soc. Biol. *27*:194–198.

Zatz, M, Frota-Pessoa, O, Levy, JA, and Peres, CA (1976) Creative phosphokinase (CPK) activity in relatives of patients with X-linked muscular dystrophies: A Brazilian study. J. Genet. Hum. *24*:153–168.

Zatz, M, Lange, K, and Spence, MA (1977) Frequency of Duchenne muscular dystrophy carriers. Lancet *1*:759.

AMERICAN JOURNAL OF PHYSICAL ANTHROPOLOGY 62:67–70 (1983)

Genetic Epidemiology

D.F. ROBERTS
Department of Human Genetics, University of Newcastle upon Tyne, Newcastle upon Tyne NE2 4AA, United Kingdom

KEY WORDS Genetic epidemiology, Multiple sclerosis

ABSTRACT From the papers in this symposium, an attempt is made to establish the scope and aim of genetic epidemiology. Specifically, its objective is seen as the elucidation of the role of genetic factors in the etiology of a disease whose distribution is related to individual genetic constitution and population genetic structure. A study of multiple sclerosis in the Orkney Islands provides an example.

Epidemiology may be defined as the study of the distribution of disease in space and time; its object is to elucidate etiological factors. Space and time, of course, refer not only to locality but also to populations and subdivisions of population, and it is in its emphasis on population variations in incidence and occurrence (essentially conceptual) that epidemiology contrasts with clinical investigations, which draws inferences from the facts observed in examination of individual patients by sight, touch, sound, and smell.

Epidemiology may be said to have been born with the work of Dr. William Farr, who was assigned responsibility for medical statistics in the Office of the Registrar General for England and Wales in 1839. The analyses embodied in the annual reports over the following 40 years, including topics such as mortality in the metal mines of Cornwall, distribution of cholera, the consequences of emigration, and the effect of imprisonment on mortality, laid the foundation for later work in their establishment of the fundamental ideas of the population at risk, the need to consider features other than disease among the groups compared, the biases inherent in selection of subjects, ways of measuring excess risk, and the concept of attributable risk. In the early days very little was known about genetic disease. Almost all the diseases known were essentially infectious, and it was in these, and especially those associated with poor hygiene, that epidemiology made its mark. One of the earliest examples was the work of John Snow who identified the Broad Street pump as the source of an epidemic of cholera in the city of London in 1854. He did so as a result of plotting the distribution of the cases of cholera in the houses and factories in that neighbourhood, and he was able to show that among the women or workers who went to that particular pump and took their water from it there were many cases, but among those who used other sources, their own wells, or in the brewery drank beer instead of water, there were no such cases.

The papers in this symposium range widely, and so help towards an understanding of what genetic epidemiology is and is not. They can be classified into several groups. First, there are those that illustrate the classic epidemiological approach, quantitation of disease incidence in one particular population. In the interesting paper on the epidemiology of cancer by Morgan et al., the fact that the population is genetically identifiable, and that incidences are compared among subdivisions of that population, implicates a relationship to genetic constitution; so this brings the work to the borders of genetic epidemiology. But this study has not as yet delved into the distribution within those subpopulations in relation to the other genes and genotypes that occur there; that is to say to the genetic structure itself, and this clearly is a task for the future. The early classical epidemiologist did not conceive of experimental modification of any environmental factors to test his hypotheses, but today though uncommon, this is not unknown, and this experimental approach is well illustrated by Mueller and Pollitt on the effects of modification of the nutritional status of mothers on the growth and body composition of their children.

Received October 26, 1982; accepted March 16, 1983.

Besides these essentially epidemiological presentations, there are those that relate less to epidemiology but illustrate the second element on which genetic epidemiology is founded, the standard procedures of medical genetic investigation and their derivatives. These are not really genetic epidemiology, for they pay little attention to the location or structure of the population, the problem at issue being the genetic basis of a particular condition. The excellent presentation by Dr. Rao and his colleagues on multifactorial inheritance of total cholesterol and triglyceride levels does, however, tend towards genetic epidemiology in its comparison of several genetically different populations. Along the same lines, essentially genetic analysis, is the excellent presentation by O'Rourke et al on manic-depressive illness, suggesting that a simple multifactorial threshold model is inadequate as an explanation of the bipolar affective disorder.

Another small group of papers, essentially genetic but this time peripherally epidemiological, includes that by Yokoyama, who uses epidemiological data on family size and marriage status in an attempt to show the effect on fitness of normal individuals through the social response to the presence of mental retardation in a family. This is a problem that exercises all who work in the genetic advisory field, well aware of the worry that comes to a couple with the knowledge that one or the other has an affected relative. Yokoyama gives an idea of how the evolutionary effect of this response can be assessed, but if further studies are to be done on the evolutionary impact of such a phenomenon, time is of the essence; for people in the west, where the best data are to be found, no longer accept the number of offspring that nature wills. They try to plan their families, taking into account all the relevant factors that are available to them, with consequent damping of fertility differentials. It would indeed be interesting to see how the figures that Yokoyama analyzed, which date for the second decade of the 20th century, compare with recent data. In this category also, similarly using family data to discuss gene frequency maintenance, is the paper of Suarez and Pierce, who effectively discount involvement of the mechanism of segregation distortion in the maintenance of the mechanism of the protease inhibitor alleles.

Finally, there is the paper by Jorde et al. who take a particularly unpleasant group of conditions, the neural tube defects, that occurred in Utah from 1940 to 1979. They carried out first a straightforward epidemiological analysis, with regard to prevalence rates, secondary sex ratios, seasonality, yearly rates, and time–space clustering. But then they turned the study into genetic epidemiology by linking their material to the Utah genealogies, so that they could examine the effect of inbreeding and, by the genealogical index method, could demonstrate substantial familial clustering of the disease. Thus they used the genetic structure of the population to provide clues to the etiological factors in these disorders. This to my mind is genetic epidemiology.

That is to say that in genetic epidemiology a further group of variables is added to the space–time–disease equation, the concepts of genetic constitution and genetic structure of the population. The questions now being asked are not, "How does a particular disease vary in occurrence in space and time?" but instead, "How do the differing frequencies of a disease between populations relate to their differences in genetic constitution?", "How does a particular disease vary in relation to the distribution of genes and genotypes within the population; that is to say with the genetic structure of that population?" The specific object is to elucidate the role of genetic factors in the etiology.

AN EXAMPLE OF GENETIC EPIDEMIOLOGY

Diseases can be regarded as occurring in a spectrum. At one end of the spectrum are those disorders that are totally genetic, which occur no matter what environment an individual grows up in. At the other end of the spectrum there are those disorders that are totally independent of the genetic constitution of the individual. There is little point in carrying out an epidemiological analysis of those disorders at the totally genetic end of the spectrum, for all that that would tell us is where individuals with the disorder live, where the family clusters, and so is really an index of past migration and settlement. The classical epidemiological approach has been particularly useful at the infectious end, the nongenetic end, of the spectrum. But the real contribution of genetic epidemiology comes in those disorders from the middle of the spectrum where it is not known how extensive is the genetic contribution to the etiology of the disorder.

Such a situation occurs in multiple sclerosis. This disease reaches its maximal incidence in the islands north of Scotland—Orkney and Shetland. Moreover, the incidence figures from the four surveys of 1954, 1962, 1970, and 1974 show a rapid rise. Rises of this degree of steepness do not occur with simple genetic disorders, unless there has been an influx of an immigrant group bringing a disease into an established population, and this clearly is not the case in Orkney and Shetland. Instead, it appears to be due to differential survival; that, as a result of modern chemotherapy, people with the disorder do not succumb to lethal infections and so survive longer, while new cases are developing at the same rate as in the past.

With multiple sclerosis there are reports in the literature of the disorder occurring in several members of the family. Some of these reports are unfortunately written by people inexperienced in clinical genetics, who regard as irrelevant and therefore ignore or are unaware of, the many unaffected members in the family. From every case of multiple sclerosis alive in Orkney on December 1, 1974, a detailed family history was obtained. The pedigrees proved to be remarkably empty of cases, no matter what mode of inheritance one envisaged as involving any major gene. The cases were then examined in relation to the genetic structure of the population (Roberts et al., 1979). For each case and for each control, the inbreeding coefficient was calculated, because if a disorder has a recessive component then the distribution of the inbreeding coefficients will be displaced towards the right. Similarly, the kinship coefficients were calculated between all possible pairs of patients, between all possible pairs of controls, and between all patients and all controls, because if a disorder has an appreciable dominant or codominant or polygenic element in its etiology, the distribution of kinship coefficients among the affecteds would be expected to shift towards the right. Incidentally, in Jorde's paper such a shift was observed in the kinship coefficients between the patients with neural tube defects. The Orkney measures of genetic structure were calculated on the basis of record searching of each case tracing the antecedents over a period of at least 200 years; that is to say to before 1776 for all cases, and indeed for many even before 1750. For every patient a contiguous control of the same sex was obtained who was born in the same parish in the same year and who had lived in that parish for the first 15 years of his life, and also a discontiguous control of the same sex, born in a discontiguous parish. By examining these two sets of controls it was possible to take into account the effect of local environment. The results showed that the mean inbreeding coefficient in the patients was high, so too were those in the controls, so the consanguinity of the patients' parents was a function of the population, not the disease. Similarly, while kinship was high among the patients, it was no higher than in controls or between patients and controls.

By this type of analysis clues can be obtained as to the presence and efficacity of major gene or polygenic effects. The next step, as Jorde showed, was to take one particular model and examine the likelihood of its occurrence. Unfortunately, with the Orkney data this has not yet been done, because none of the families has a sufficient number of affected individuals in the family for the models to be tested. But with the passage of time, one expects that further data will accumulate and it may then be possible to do so.

However, this analysis, and our quite massive survey of other genetic characters, led us to think that, if there is a genetic component to this condition, it is likely to be polygenic. Heritability was therefore calculated. A particularly interesting finding was that whereas the heritability of multiple sclerosis in the Newcastle region is about 52%, the heritability in Orkney is significantly lower at about 31%. These parallel studies in different localities indicate what genetic theory has always emphasised, that estimates of heritability only relate to a particular population in a particular environment at a particular time. Heritability calculated in one population cannot be applied for the same disorder in some other population elsewhere.

The demonstration of these differences between populations is extremely important, for one practical application of any heritability estimate is in the estimation of risks to relatives. Such risks would also be expected to differ, and this again is demonstrated by reference to Jorde's paper. He noted that where one child had been born with a neural tube defect, the recurrence risk that a sib would also be born with a defect was 1 in 30. In Newcastle, this risk is about 1 in 18, an appreciable difference.

CONCLUSIONS

There is much to be gained from defining genetic epidemiology as done here, and this can be seen in the paper by Jorde on neural tube defects and in our own on multiple sclerosis in Orkney. Yet one would hope that such analyses may come to be regarded as studies of historical rather than of future health. This is particularly the case with Jorde's paper. The neural tube defects are disorders that are almost completely avoidable. Over 98% can be avoided by simple blood screening followed by the more complex prodcedure of amniocentesis to confirm a positive result, followed, if this again proves positive, by termination of the pregnancy. If such screening methods result in a massive decrease in incidence of neural tube defects, then a study such as Jorde's would not be possible. If other disorders, with our increasing knowledge of how to treat or avoid them, are also on the decrease, then perhaps if the methods of genetic epidemiology here outlined are to be applied in an attempt to understand the factors responsible for them, this task should be initiated fairly quickly. One only has to look at the history of infectious diseases in childhood. No, or at least very few, children today in the Western world die of whooping cough, measles, typhus, typhoid, and other old killing diseases of childhood. Yet it is only a century ago that at least one in three children died from such disorders. If that has happened with infectious disease, then it is not overoptimistic to expect that within the foreseeable future there will be further change in the disease pattern. For genetic epidemiology to realize its full contribution to the understanding of disease, its application must be regarded as a matter of urgency.

LITERATURE CITED

Roberts, DF, Roberts, MJ, and Poskanzer, DC (1979) Genetic analysis of multiple sclerosis in Orkney. J. Epidemiol. Commun. Hlth *33*:236–242.

AMERICAN JOURNAL OF PHYSICAL ANTHROPOLOGY 62:71–79 (1983)

Genetic and Evolutionary Implications in Peptic Ulcer Disease

GLORIA M. PETERSEN AND JEROME I. ROTTER
Division of Medical Genetics, Departments of Medicine and Pediatrics, UCLA School of Medicine, Harbor-UCLA Medical Center, Torrance, California 90509

KEY WORDS Peptic ulcer disease, Genetic heterogeneity

ABSTRACT The evidence for a genetic component in peptic ulcer disease has been based on twin, family, and blood group studies. A polygenic model for the inheritance of peptic ulcers has been displaced by a genetic heterogeneity model based on several lines of evidence, some of the most powerful being recent work using subclinical markers. One marker in particular, an elevated level of serum pepsinogen I (PG I), a pepsin precursor produced by the gastric mucosa, secreted into the stomach lumen and also appearing in the bloodstream, has been found to be associated with a subgroup of duodenal ulcer patients. Segregation analysis of elevated serum PG I in duodenal ulcer sibships demonstrates familial aggregation consistent with autosomal dominant inheritance. Elevated PG I is also accompanied by gastric hyperacidity and presumably indicates those individuals with an increased mass of chief and parietal cells, and thus an increased capacity for peptic activity, an important element in the pathogenesis of ulcer disease. An evolutionary hypothesis based on selection for peptic activity and acidity is offered to explain several of the epidemiologic and genetic elements of this group of chronic diseases.

A peptic ulcer is a hole in the lining of the gastrointestinal tract in those areas exposed to acid and pepsin. The most common sites are the stomach (gastric ulcer), or the first part of the small intestine, the duodenum (duodenal ulcer). A peptic ulcer occurs when the processes of mucosal defense and regeneration are overwhelmed by the aggressive activity of gastric acid and pepsin. Althouth virtually unknown prior to the latter half of the nineteenth century (Jennings, 1940; Ivy et al., 1950; Wylie, 1981), peptic ulcer disease is now considered among the commonest of chronic diseases with a lifetime prevalence of some 8–10% in European and North American populations (Grossman, 1980).

The physiology, diagnosis, and therapy of peptic ulcer have been studied extensively by many investigators, but the underlying genetic basis for this disorder has received less attention and is still incompletely understood. Hereditary factors are an important component in peptic ulcer disease susceptibility (McConnell, 1980; Rotter, 1980b); however, it has been difficult to analyze this genetic contribution in that peptic ulcer disease is a relatively common, chronic disorder, with variable diagnostic criteria. Evidence now supports the hypothesis that this disease, or more accurately, group of diseases, has multiple genetic and environmental etiologies. We will review this evidence and suggest an explanatory model for the increase of peptic ulcer during the last century based on selection for increased peptic activity/gastric acidity through infectious disease agents, in the process reconciling several historic, epidemiologic, and genetic observations.

GENETICS OF PEPTIC ULCER

The earliest evidence suggesting a genetic component in ulcer disease is based on analyses of families, twins, and blood group associations. A series of family studies showed that first-degree relatives of index cases with ulcers were at an increased risk for developing ulcer disease when compared to appropriately age-matched controls (McConnell, 1966; Rotter, 1980b). In addition, Doll and Kellock (1951) showed that relatives of a gastric ulcer patient were at an increased risk of developing gastric

Received August 5, 1982; accepted March 16, 1983.

ulcer but not duodenal ulcer, and similarly, relatives of a duodenal ulcer patient were at an increased risk of developing duodenal ulcer but not gastric ulcer (Table 1); these observations suggested that these were two separately inherited entities.

Twin studies also indicated a genetic factor: Although the concordance for peptic ulcer in monozygotic twins is not 100%, it is consistently higher when compared to dizygotic twins (Table 2). In a follow-up to the Danish registry twin study by Harvald and Hauge (1958), Gotlieb-Jensen (1972) found that the ulcer site was also concordant when both twins were affected (both had gastric ulcer or both had duodenal ulcer), further supporting the concept that separate genetic factors predisposed to each disorder.

Finally, the association of duodenal ulcer with genetic polymorphisms, specifically, blood group 0 and nonsecretor status, has been well established by numerous studies, but the physiologic relationship remains unknown, probably because of the low order of magnitude of the association (Mourant et al., 1978; McConnell, 1980).

These early lines of evidence supported the existence of a genetic factor or factors for peptic ulcer disease, but the observed familial aggregation was found to be inconsistent with any simple Mendelian forms of inheritance. Thus, as it was difficult to fit the data to a monogenic model, the mode of inheritance had been presumed to be polygenic: many genes of small effect additively working together to produce the ulcer phenotype (McConnell, 1966; Cowan, 1973).

Although the polygenic model is consistent with the blood group data, it lacks adequate explanatory power. Rotter and Rimoin proposed an alternative: that peptic ulcer disease is genetically heterogeneous (Rotter et al., 1976). They posited that peptic ulcer is a group of distinct diseases with different genetic and nongenetic etiologies which share a clinical phenotype. Thus, a search for potential genetic variation in *subclinical markers* (underlying physiologic or biochemical abnormalities) might provide a means of delineating different diseases. This strategy for elucidating genetic heterogeneity would thus identify the different disorders, which otherwise in the aggregate might appear to mimic the polygenic model. In the process, one might clarify the potentially simpler modes of inheritance of the individual diseases (Rotter and Rimoin, 1977). The evidence supporting genetic heterogeneity of peptic ulcer disease now encompasses numerous lines of evidence. This includes the independent distribution among families of gastric and duodenal ulcer (Doll and Kellock, 1951; Gotlieb-Jensen, 1972), and ethnic variation in incidence, type of ulcer, and clinical characteristics (Lam et al., 1980; Tovey, 1979). Peptic ulcer is also involved in over ten rare genetic syndromes (Rotter and Rimoin, 1981), such as the autosomal dominantly inherited multiple endocrine adenoma syndrome, Type I (Lamers et al., 1978). Although there are several methods of identifying genetic heterogeneity (Rotter, 1980a; 1981), we will emphasize one approach herein, the genetic analysis of subclinical markers.

SUBCLINICAL MARKERS

Subclinical markers are parameters used to detect the abnormal genotype in the absence of the full disease phenotype (i.e., abnormal glucose tolerance in diabetes, decreased transferrin saturation levels in hemochromatosis). These markers represent abnormalities directly involved in the pathogenesis of the disease. They are useful in genetic studies because in many disorders not all individuals with the mutant genotype may manifest the disorder (reduced penetrance), the variability of the

TABLE 1. Excess (EXC) of ulcer types between observed (OBS) and expected (EXP) in first degree relatives of peptic ulcer probands as a function of ulcer location[1]

	Type of ulcer in relatives											
	Gastric						Duodenal					
	Males			Females			Males			Females		
Type of ulcer in proband	OBS	EXP	EXC	OBS	EXP	EXC	OBS	EXP	EXC	OBS	EXP	EXC
Gastric	21	6.8	14.2	9	3.6	5.4	13	12.1	0.9	5	2.2	2.8
Duodenal	10	6.5	3.5	3	3.1	−0.1	42	13.5	28.5	6	2.0	4.0

[1]Adapted from Doll and Kellock (1951).

TABLE 2. Peptic ulcer in twins

No. of pairs studied	Concordance (%) MZ	DZ	Reference
10	50	12.5	Doig (1957)
181	18	7.3	Harvald and Hauge (1958)
112	50	14.1	Eberhard (1968)
837	11.3	5.8	Pollin et al. (1969)[1]
167	52.6	35.7	Gotlieb-Jensen (1972)[2]

[1]Duodenal ulcer only.
[2]Followup of Harvald and Hauge study.

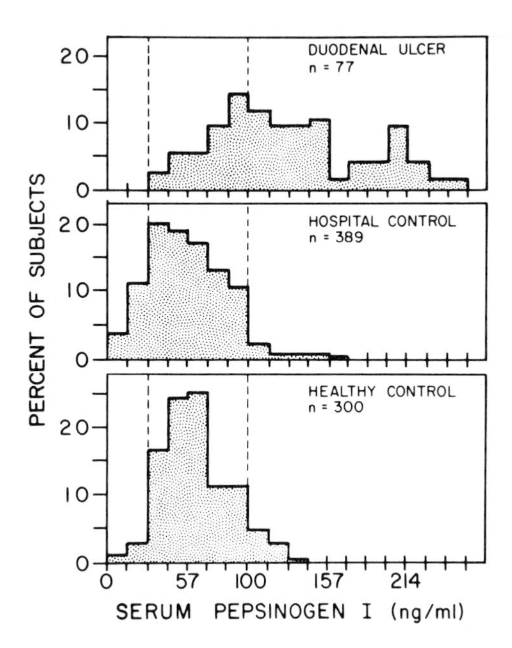

Fig. 1. Distribution of serum pepsinogen I in duodenal ulcer patients and control groups. Vertical dashed lines indicate normal limits. Adapted from Samloff et al. (1975a).

phenotype may be so great that the clinical features are too mild to be readily apparent (variable expressivity), or there may be a delayed age of onset of the disease such that the younger genetically predisposed individuals would be clinically normal. Thus, subclinical markers maximize the number of affected individuals that can be detected. By appropriate genetic-physiologic studies and by stratifying ulcer patients into groups according to these different markers (Table 3), distinct ulcer syndromes can be defined (Rotter, 1980b).

Because of its potential as a subclinical marker, we have conducted extensive genetic studies of pepsinogen I in duodenal ulcer patients and their families. Pepsinogen is the precursor to pepsin, the proteolytic enzyme responsible for the digestion occurring in the stomach lumen. It is produced by the oxyntic glands located in the lining of the stomach and duodenum and is activated by acid. It is a heterogeneous protein, being composed of two immunologically distinct groups, pepsinogen I (PG I) and pepsinogen II (PG II) (Samloff, 1969, 1982). These groups differ electrophoretically—PG I migrates more anodally than PG II; in cellular origin—PG I is produced only in the stomach fundus, while PG II is produced in stomach and duodenum; and in distribution in body fluids—PG I is found in serum and urine while PG II is found in serum and seminal fluid (Samloff, 1979). Studies of PG I and PG II levels in serum by radioimmunoassay have shown that the PG I level is related to peptic cell mass and acid secretory capacity of the stomach (Samloff et al., 1975b). Further, a survey of duodenal ulcer patients and healthy controls showed that PG I levels in duodenal ulcer patients were bimodally distributed, with more than half of the patients having higher levels than controls (Fig. 1) (Samloff et al., 1975a).

Analysis of the distribution of serum PG I in two large duodenal ulcer pedigrees suggested that elevated levels were transmitted in an autosomal dominant fashion (Rotter et al., 1979a). We investigated this further in a genetic analysis of this trait in 123 sibships ascertained through duodenal ulcer at the Broadgreen Hospital in Liverpool, England (Rotter et al., 1979b). The sibships could be divided into those where the ulcer patients and their sibs had normal PG I levels (normo PG I) and those where the ulcer patient had an elevated PG I level (hyper PG I). The mean PG I levels of the 57 nonulcer sibs in the hyper PG I group (91.2 ± 52.1 ng/ml) was found to be significantly higher than in the 35 nonulcer sibs of the normo PG I group (63.1 ± 33.0 ng/ml) ($p < .001$), and intermediate to the mean PG I level in the 49 unoperated hyper PG I ulcer patients (133.5 ± 51.4 ng/ml). In contrast, the PG I levels of sibs of the normo PG I nonulcer patients did not differ significantly from healthy controls. Analysis of extended pedigrees yields similar results (Rotter et al., 1981).

TABLE 3. Subclinical markers for peptic ulcer disease[1]

Abnormalities of acid and gastrin secretion
*Increased maximal acid output
Increased acid response to a meal
Increased sensitivity of acid stimulation by gastrin
Decreased inhibition of acid secretion by acid or distention
Decreased inhibition of gastrin release by acid
*Exaggerated serum gastrin response to a meal
*Increased rate of gastric emptying
Increased mass of chief (pepsinogen-producing) and parietal (acid-producing) cells.
*Elevated serum pepsinogen I level
Calcium transport
Increased calcium absorption and renal transport
Increased serum catecholamines
Gastritis/duodenitis
Low serum pepsinogen I/pepsinogen II ratio
Decreased alpha-1 antitrypsin levels
Immunologically mediated forms of ulcer
Acid stimulated antibodies
Antibodies to secretory IgA
Secretin induced histamine release in skin
Smooth muscle contracting factor
Acetylcholinesterase levels in serum and red cells
Salivary response to citric acid
Duodenogastric reflux and pyloric sphincter dysfunction

*, confirmed in genetic studies.
[1]Updated from Rotter (1980a), Rotter and Heiner (1982).

Thus, our initial analyses indicated we were dealing with at least two genetic forms of duodenal ulcer whose separation, based on the pepsinogen I subclinical marker, divides them into hyper PG I and normo PG I groups. We then performed segregation analysis of the more homogeneous hyper PG I group, composed of 47 sibships characterized by an index ulcer patient with PG I > 100 ng/ml.

Segregation analysis is a commonly used genetic method which tests whether a trait is inherited in a hypothesized Mendelian manner. A segregation ratio is calculated from observed data and is then compared to the ratio predicted by specific genetic hypotheses. For example, the predicted ratio is 0.25 if a trait is inherited in an autosomal recessive fashion with both parents unaffected. Similarly, for an autosomal dominant trait with one affected parent, the predicted segregation ratio is 0.5.

The methods of segregation analysis vary with the different modes in which the families being studied were ascertained. Single ascertainment assumes that each sibship is ascertained through one affected individual (proband) regardless of the number of affected offspring in the family. Analyses making this assumption thus give a lower bound for the segregation ratio since they assume a lower probability of ascertainment than the other methods. Complete truncate ascertainment assumes that every affected individual in the community has been ascertained regardless of whether or not they have affected relatives. Although in practice this is not the usual case, analyses making this assumption give an upper bound for the segregation ratio. Multiple incomplete ascertainment assumes that there may be more than one proband per sibship, and is the most likely ascertainment scheme. Segregation analyses using this ascertainment scheme give the more realistic estimates of the segregation ratio (Morton, 1976; Emery, 1976).

Our analysis was complicated by the fact that we had ascertained through duodenal ulcer disease, but were testing elevated PG I as the trait of interest. We thus repeated the analysis using combinations of each of the above assumptions and each of two classification schemes where (1) the affected phenotype was defined as elevated PG I alone, and (2) the affected phenotype was defined as duodenal ulcer and/or elevated PG I. The latter was necessitated by the fact that a number of patients had undergone acid-reducing, and hence PG I reducing, operations prior to their entry into the study. The results of this analysis (Table 4) indicated that the segregation ratios calculated from these data were most consistent with an autosomal dominant hypothesis and reject the autosomal recessive hypothesis. Under the most realistic ascertainment scheme (multiple

TABLE 4. Segregation ratios (±S.E.) of elevated serum pepsinogen I (PG I) in duodenal ulcer (DU) sibships[1]

	Definition of affected phenotype	
	Elevated PG I	Elevated PG I or DU, or both
Mode of ascertainment:		
Complete truncate	0.46 ± 0.06	0.60 ± 0.05
Multiple incomplete	0.34 ± 0.06[2]	0.55 ± 0.05[3]
Single incomplete	0.32 ± 0.06	0.46 ± 0.05

[1]Adapted from Rotter et al. (1979b).
[2]Ascertainment probability estimated at 0.22 ± 0.10.
[3]Ascertainment probability estimated at 0.67 ± 0.07.

incomplete), with affected phenotype defined as duodenal ulcer and/or hyper PG I, the segregation ratio was 0.55 ± 0.05.

Overall, the analyses demonstrated that there are at least two subtypes of duodenal ulcer: (1) a hyper PG I form where a genetic predisposition to duodenal ulcer in apparently normal sibs can be identified by elevated serum PG I levels and where the aggregation of elevated PG I is most consistent with dominant inheritance, and (2) a normo PG I form where the ulcer patient and his sibs have normal PG I levels. Additional lines of evidence suggest the existence of further subgroups based on other subclinical markers such as the gastrin response to a protein meal or the rate of gastric emptying (Rotter, 1980b). The accumulating evidence supports the notion that rather than resulting from entirely environmental factors, ulcer disease may have a strong yet heterogeneous genetic component, consisting of multiple disorders, each with potentially different genetic predispositions.

AN EVOLUTIONARY HYPOTHESIS

An interesting problem then, is explaining the effect of evolutionary forces on a genetically heterogeneous disease with a common, variable, and generally nonlethal phenotype. From this perspective, the question arises as to how such a disorder rarely observed prior to the last half of the nineteenth century could have increased in frequency to that now observed in Western populations (Jennings, 1940; Wylie, 1981). Changes in diagnostic criteria cannot fully explain this trend (Ivy et al., 1950). Environmental and psychological explanations have been offered (i.e., Susser, 1967); but they may require reevaluation in light of the recent genetic and physiologic findings.

We propose one possibility: that at least the hypersecretor forms of peptic ulcer disease in Western populations may be a consequence of selection by infectious bacterial disease agents (such as tuberculosis) against individuals with low gastric acidity, during the preceding century. The result is a greater frequency of individuals with higher gastric proteolytic activity, with the untoward effect of increased peptic disease prevalence. The notion that infectious disease has an evolutionary impact on genetic traits has been proposed by others (Livingstone, 1960; Motulsky, 1960); the malaria and sickle cell gene model is now a classic example of this interaction (Allison, 1954, 1964). However, for common diseases with less well-defined genetic predispositions, an evolutionary relationship is not as apparent. Our hypothesis draws from the epidemiology and history of peptic ulcer disease and tuberculosis (TB), their pathophysiologies, and their relationship to gastric acidity.

An examination of the epidemiology and temporal changes in infectious disease of bacterial origin reveals that they have declined significantly over the past century (Burnet and White, 1972). It is presumed that improved living conditions, better hygiene, and medical advances in treatment have contributed to these shifts. However, it is also possible that genetic factors may be involved, in that selection for less susceptible individuals may have occurred. For example, TB mortality began declining in England and Wales by 1850, before the tubercle bacillus was identifed, and before therapeutic or public health measures had been developed (Kass, 1971; Stead and Bates, 1980), though it continued to be a leading cause of death in Europe throughout the latter part of the nineteenth century (Clarke, 1952; McDougall, 1949; Blacklock, 1947).

A genetic factor for resistance to TB had long been suspected and early studies had found this to be a strong likelihood (Wright and Lewis, 1921; Kallman and Reisner, 1943; Clarke, 1952), though the precise mechanism was unknown. It is now likely that in many cases one can reasonably implicate the immune response, which is a specific defensive factor (Youmans, 1979). We suggest in addition, that gastric acidity, a reflection of peptic cell mass, is a *non-specific* defense mechanism which also may have been subject to selection pressure. The intensity of pressure effected by TB, with an annual peak mortality rate by 1830 of 300/100,000 in England and Wales, and an even higher morbidity rate (Clarke, 1952), would have influenced the frequency of those traits which would enable survival and reproduction of offspring. Thus, the gradual decline of TB during the latter part of the nineteenth century was undoubtedly due to several causes, but among these was likely an increased resistance to infection conferred in the survivors by a variety of factors, one of which may have been higher gastric acidity (see below). With the institution of extensive public health measures and chemotherapy, the selective pressures have been relaxed. Importantly, concurrent with the decline in TB mortality, Western populations experienced an increasing prevalence of peptic ulcer disease from an absent or rare disorder, to the present rate of 8–10%, and a corresponding increase in mortality due to peptic ulcer (Fig. 2) (Ivy et al., 1950; Langman, 1979). It is of interest that the frequency of this disorder (as opposed to its mortality) may not have declined over the last few decades (Kurata et al., 1983).

A potential mechanism for this epidemiologic relationship could lie in the physiology of resistence to tubercular infection. Human TB develops by inhalation of droplet nuclei in the air containing the tubercle bacilli, which then invade lymphatic and pulmonary tissues. Reinfection within the individual can occur as infected sputum is coughed up and swallowed, exposing the tissue to a further bacterial load. Bovine TB in humans occurs when infected milk is ingested. The gastrointestinal tract may thus play a role in the mediation of TB infection. We suggest that a property of the gastrointestinal tract, gastric acidity, was an important evolutionary factor in human susceptibility to bacterial infections. Our studies have demonstrated that there is a genetic basis for the variation in gastric proteolytic activity, which reflects peptic cell mass and gastric secretory capacity. Even in normal individuals there appears to be a large genetic component to pepsinogen I levels (Rotter et al., 1982). It has long been established that gastric acid has a bacteriocidal effect (Arnold, 1933; Gray and Shiner, 1967) Achlorhydric persons do not have the sterile stomachs, as found in normal individuals (Drasar et al., 1969). This "gastric acid barrier" to ingested bacteria would appear to be potentially clinically important, as persons with low-acid stomachs are more susceptible to infection (Giannella et al., 1972, 1973). Experiments by Nalin et al. (1978) demonstrated that hypochlorhydria was a prerequisite for and *preceded* cholera vibrio infection, rather than resulting from the infection itself. We suggest that an analogous situation for TB and hypochlorhydria may have existed during the endemic and epidemic periods in Western populations, for which detailed TB epidemiologic data are available.

There are several observations which lend support to this hypothesis. First, studies of gastric acid production reveal that the periods of low acid production, less than 2 years and greater than 50 years (Vanzant et al., 1932;

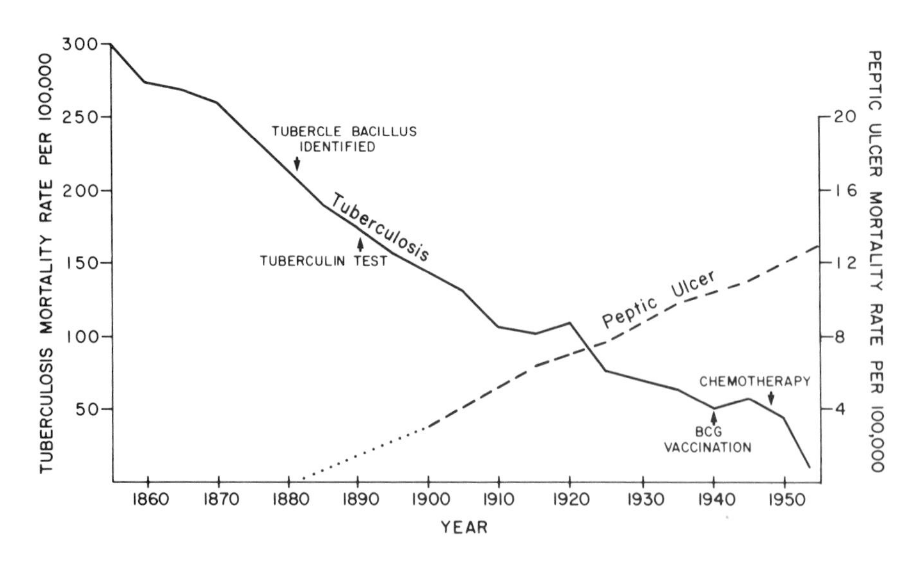

Fig. 2. Mortality trends of tuberculosis and peptic ulcer from data for England and Wales. Adapted from Kass (1971) and Ivy et al. (1950).

Levin et al., 1951; Yamaguchi et al., 1968), are congruent with the ages most susceptible to TB infection and mortality (Clarke, 1952). Second, gastric acidity measured in TB patients in early studies revealed that those with low acid to no acid comprised an important segment of the TB subjects, ranging from 15% (Tørning, 1929) and 30% (Brooke, 1928), to 63% (Rave, 1930) and 75% (Perla, 1926). This wide range was probably due to technical variation, but all studies were consistent in the observation that the frequency of hyperacidity in TB patients was lower than in non-TB controls. Third, a possible contradiction to our hypothesis is the purported positive clinical association of TB with peptic ulcer, i.e., that the two occur together more frequently than expected from their individual disease frequencies (Sturtevant and Shapiro, 1931; Langman and Cooke, 1976). Reasoning from our hypothesis, if hyperacidity were protective, then a negative association would be expected. A detailed examination of such studies reveals that in fact, peptic ulcer patients have an increased susceptibility to TB only following gastrectomy, a surgical procedure which removes the acid-producing cell mass of the stomach (Steel and Johnston, 1956; Sturtevant and Shapiro, 1931; Viskum, 1974; Langman and Cooke, 1976) Postoperative weight loss, stress, and other surgical sequelae could not be implicated in this heightened susceptibility to TB infection, as this relationship appeared to be independent of such factors (Boman, 1955 1956; Balint, 1958). Bacterial growth is also increased in the resected stomach (Schwabacher et al., 1959; Bach-Nielsen and Amdrup, 1965). These clinical studies support the existence of a protective role for gastric acidity with respect to tubercular disease.

Thus, gastric acidity is a potentially critical mechanism in providing a nonspecific defense against bacterial infection. With repeated exposure or acute exposure to ingested infectious bacterial agents, individuals with lower gastric acidity may have been selected against, such that there may have been a shift to higher population mean gastric acidity and proteolytic activity. Under the intense selective pressures both by epidemic waves of TB and enteric bacterial infections and by chronic endemic infection in the crowded living conditions of the newly industrialized Western world, it is conceivable that a higher gastric acid phenotype may have been advantageous. In the susceptible organism or individual, TB can be virulent and result in rapid mortality; as this disease most often affected individuals prior to their reproductive period, it could have had a major selective effect.

A consequence of having the higher gastric acid phenotype would be a heightened defense against infectious diseases mediated through oral gastric exposure, but would also have led to an increased predisposition to acid-peptic disease, with the eventual endpoint being peptic ulcer seen in Western populations. Our studies and those of others investigators are revealing that a variety of mechanisms can produce a high gastric acid-peptic activity phenotype and that many of these mechanisms have an inherited basis. Peptic ulcer disease, a chronic but usually nonlethal disorder, is thus genetically and pathogenetically heterogeneous. The historical and epidemiologic data suggest that selection against lower gastric acidity would have needed three to four generations to manifest an effect in increased peptic ulcer disease. This period of change is feasible, given the magnitude of selection and the heterogeneous etiologies of higher gastric acidity (Petersen and Rotter, in preparation).

Thus, we suggest that the evolution of hypersecretory forms of peptic ulcer disease may be a consequence of selection by infectious disease agents for higher gastric proteolytic activity. This hypothesis accounts for the historic, epidemiologic, clinical, and physiologic observations. Most importantly, it incorporates the recent findings that peptic ulcer disease is extensively genetically heterogeneous, in that a variety of inherited subclinical markers predispose to a number of different diseases whose common endpoint is a shared phenotype. This is particularly so for the hypersecretory forms of ulcer. In addition, other mechanisms, such as gastrointestinal immunity, are likely to have played a role in resistance to a host of diseases whose infective routes involved the gastrointestinal tract. It is possible that overactivity of such mechanisms may result in other, often normosecretor, forms of peptic ulcer, on an immunologic basis (Rotter and Heiner, 1982). A combination of these influences could well have made a major contribution to the appearance of the many disorders that in the aggregate we call today peptic ulcer disease.

ACKNOWLEDGMENTS

This research was supported by a center for Ulcer Research and Education training grant, AM07180, a National Institutes of Health Individual Fellowship, NIAMDD 5F32AM06362, and NIADDK grant, AM17328.

LITERATURE CITED

Allison, AC (1954) Protection by the sickle-cell trait against subtertian malarial infection. Br. Med. J. *1:*290.

Allison, AC (1964) Polymorphism and natural selection in human populations. Cold Spring Harbor Symp. Quant. Biol. *29:*137–149.

Arnold, L (1933) The bacterial flora within the stomach and small intestine. Am. J. Med. Sci. *186:*471–481.

Bach-Nielsen, P, and Amdrup, E (1965) Peroperative bacteriologic examination of the stomach and duodenum. Acta Chir. Scand. *129:*521–529.

Balint, JA (1958) Pulmonary tuberculosis and partial gastrectomy. Gastroenterologia *90:*65–84.

Blacklock, JWS (1947) The epidemiology of tuberculosis. Br. Med. J. *1:*707–712.

Boman, K (1955–56) Tuberculosis occurring after gastrectomy. Acta Chir. Scand. *110:*451–457.

Brooke, COSB (1928) Gastric secretion in phthisis. Lancet *2:*1128–1129.

Burnet, M, and White, DO (1972) Natural History of Infectious Disease. Cambridge: Cambridge U. Press.

Clarke, BR (1952) Causes and Prevention of Tuberculosis. Edinburgh: Livingstone.

Cowan, WK (1973) Genetics of duodenal and gastric ulcer. Clin. Gastroenterol. *2:*539–545.

Doig, RK (1957) Illness in twins: Duodenal ulcer. Med. J. Aust. *2:*617–619.

Doll, R, and Kellock, TD (1951) The separate inheritance of gastric and duodenal ulcers. Ann. Eugen. *16:*231–240.

Drasar, BS, Shiner, M and McLeod, GM (1969) Studies on the intestinal flora. Gastroenterology *56:*71–79.

Eberhard, G (1968) Peptic ulcer in twins. Acta Psychol. Scand. [Suppl.] *205:*4–118.

Emery, AEH (1976) Methodology in Medical Genetics. New York: Churchill Livingstone.

Giannella, RA, Broitman, SA, and Zamcheck, N (1972) Gastric acid barrier to ingested microorganisms in man: studies *in vivo* and *in vitro*. Gut *13:*251–256.

Giannella, RA, Broitman, SA, and Zamcheck, N (1973) Influence of gastric acidity on bacterial and parasitic enteric infections. Ann. Int. Med. *78:*271–276.

Gotlieb-Jensen, K (1972) Peptic Ulcer: Genetic and Epidemiologic Aspects Based on Twin Studies. Copenhagen: Munksgaard.

Gray, JDA, and Shiner, M (1967) Influence of gastric pH on gastric and jejunal flora. Gut *8:*574–581.

Grossman, MI (1980) Peptic ulcer: Definition and epidemiology. In JI Rotter, IM Samloff, and DL Rimoin (eds): Genetics and Heterogeneity of Common Gastrontestinal Disorders. New York: Academic Press, pp. 21–29.

Harvald, B, and Hauge, M (1958) A catamnestic investigation of Danish twins. Acta Genet. *8:*287–294.

Ivy, AC, Grossman, MI, and Bachrach, WH (1950) Peptic Ulcer. Philadelphia: Blakiston.

Jennings, D (1940) Perforated peptic ulcer. Lancet *1:*395–398.

Kallman, FJ, and Reisner, D (1943) Twin studies on the significance of genetic factors in tuberculosis. Am. Rev. Respir. Dis. (formerly Am Rev. Tuberc) *47:*549–574.

Kass, EH (1971) Infectious diseases and social change. J. Infect. Dis. *123:*110–114.

Kurata, JH, Elashoff, JD, Haile, BM, and Honda, GD (1983) A reappraisal of time trends in ulcer disease: factors related to changes in ulcer hospitalization and mortality rates. Am. J. Public Health (in press).

Lam, SK, Hasan, M, Sircus, W, Wong, J, Ong, GB, and Prescott, RJ (1980) A comparison of the maximal acid output and gastrin response to meals in Chinese and Scots, normals and with duodenal ulcer. Gut *21:*324–328.

Lamers, CBH, Stadil, F, and Van Tongeren, JH (1978) Prevalence of endocrine abnormalities in patients with the Zollinger-Ellison syndrome in their families. Am. J. Med. *64:*607–612.

Langman, MJS (1979) The Epidemiology of Chronic Digestive Disease. London: Edward Arnold Pub.

Langman, MJS, and Cooke, AR (1976) Gastric and duodenal ulcer and their associated diseases. Lancet *1:*680–683.

Levin, E, Kirsner, JB, and Palmer, WL (1951) A simple measure of gastric secretion in man. Gastroenterology *19:*88–98.

Livingstone, FB (1960) Natural selection, disease, and ongoing human evolution, as illustrated by the ABO blood groups. Hum. Biol. *32:*17–27.

McConnell, RB (1966) The Genetics of the Gastrointestinal Disorders. London: Oxford U. Press.

McConnell, RB (1980) Peptic ulcer: Early genetic evidence. In JI Rotter, IM Samloff, and DL Rimoin (eds): Genetics and Heterogeneity of Common Gastrointestinal Disorders. New York: Academic Press, pp. 32–41.

McDougall, J (1949) Tuberculosis. Edinburgh: Livingstone.

Morton, NE (1976) Segregation analysis. In NE Morton (ed): Computer Applications in Genetics. Honolulu: U. Hawaii Press, pp. 129–140.

Motulsky, AG (1960) Metabolic polymorphisms and the role of infectious disease in human evolution. Hum. Biol. *32:*28–62.

Mourant, AE, Kopec, AC, and Domaniewska-Sobczak, K (1978) Blood Groups and Diseases. Oxford: Oxford U. Press.

Nalin, DR, Levine, RJ, Levine, MM, Hoover, D, Bergquist, E, McLaughlin, J, Libonati, J, Alam, J, and Hornick, RB (1978) Cholera, non-vibrio cholera and stomach acid. Lancet *2:*856–859.

Perla, D (1926) Studies on gastric function in pulmonary tuberculosis. Am. Rev. Respir. Dis (formerly Am Rev. Tuberc). *13:*317–326.

Pollin, W, Allen, MG, Hoffer, A, Stabenau, JR, and Hrubec, Z (1969) Psychopathology in 15,909 pairs of veteran twins. Am. J. Psychiatry *126:*597–610.

Rave, E (1930) Über den Magenchemismus bei Lungentuberkulose. Z. Klin. Med. *113:*621–640.

Rotter, JI (1980a) Genetic approaches to ulcer heterogeneity. In JI Rotter, IM Samloff, and DL Rimoin (eds): Genetics and Heterogeneity of Common Gastrointestinal Disorders. New York: Academic Press, pp. 111–128.

Rotter, JI (1980b) The genetics of peptic ulcer: More than one gene, more than one disease. In AG Steinberg, AG Bearn, AG Motulsky, and B Childs (eds): Prog. Med. Genet. *4:*1–58.

Rotter, JI (1981) Gastric and duodenal ulcer are each many different diseases. Dig. Dis. Sci. *26:*154–160.

Rotter, JI, and Heiner, DC (1982)Are there immunologic forms of duodenal ulcer? J. Clin. Lab. Immunol. *7:*1–6.

Rotter, JI, and Rimoin, DL (1977) Peptic ulcer disease–A heterogeneous group of disorders? Gastroenterology *73:*604–607.

Rotter, JI, and Rimoin, DL (1981) The genetic syndromology of peptic ulcer. Am. J. Med. Genet. *10:*315–321.

Rotter, JI, Gursky, JM, Samloff, IM, and Rimoin, DL (1976) Peptic ulcer disease further evidence for genetic heterogeneity. Excerpta Medica, Int. Cong. Ser. *397:*96.

Rotter, JI, Sones, JQ, Samloff, IM, Gursky, JM, Richardson, CT, Walsh, JH, and Rimoin, DL (1979a) Duodenal ulcer disease associated with elevated serum pepsinogen I, an inherited autosomal dominant disorder. N. Engl. J. Med. *300:*63–66.

Rotter, JI, Petersen, GM, Samloff, IM, McConnell, RB, Ellis, A, Spence, MA, and Rimoin, DL (1979b) Genetic heterogeneity of familial hyperpepsinogenemic I and normopepsinogenemic I duodenal ulcer disease. Ann. Int. Med. *91:*372–377.

Rotter, JI, Petersen, GM, Samloff, IM, Rubin, R, Apostol, R, and Rimoin, DL (1981) Variation in serum pepsinogen I in duodenal ulcer patients and their healthy relatives. Clin. Res. *29:*87A.

Rotter, JI, Wong, FL, Samloff, IM, Varis, K, Siurala, M, Ihamaki, T, Ellis, A, and McConnell, RB (1982) Evidence

for a major dominance component in the variation of serum pepsinogen I levels. Am. J. Hum. Genet. *34:*395–401.
Samloff, IM (1969) Slow moving protease and the seven pepsinogens. Gastroenterology *57:*659–669.
Samloff, IM (1979) Serum pepsinogens I and II. In JE Berk (ed): Developments in Digestive Disease. Philadelphia: Lea and Febiger, pp. 1–12.
Samloff, IM (1982) Pepsinogens I and II. Gastroenterology *82:*26–33.
Samloff, IM, Liebman, WM and Panitch, NM (1975a) Serum group I pepsinogens by radioimmunoassay in control subjects and patients with peptic ulcer. Gastroenterology *69:*83–90.
Samloff, IM, Secrist, DM, and Passaro E, Jr. (1975b) A study of the relationship between serum group I pepsinogen levels and gastric acid secretion. Gastroenterology *69:*1196–1200.
Schwabacher, H, Salsbury, AJ, and Loosemore, TGE (1959) The bacterial flora of gastric lavages from patients undergoing partial gastrectomy. J. Clin. Pathol. *12:*565–567.
Stead, WW, and Bates, JH (1980) Epidemiology and prevention of tuberculosis. In A Fishman (ed): Pulmonary Disease and Disorders. New York: McGraw-Hill, pp. 1234–1254.
Steel, SJ, and Johnston, RN (1956) Peptic ulcer and pulmonary tuberculosis. Br. J. Tuberc. *50:*233–238.
Sturtevant, M, and Shapiro, LL (1931) Peptic ulcer, association with pulmonary tuberculosis. Arch. Int. Med. *48:*1198–1202.
Susser, M (1967) Causes of peptic ulcer. J. Chronic Dis. *20:*435–456.
Tørning, K (1929) Über die Magensekretion der Phthisiker. Klin. Wochenschr *8:*1261–1263.
Tovey, FI (1979) Peptic ulcer in India and Bangladesh. Gut *20:*329–347.
Vanzant, FR, Alvarez, WC, Eusterman, GB, Dunn, HL, and Berkson, J (1932) The normal range of gastric acidity from youth to old age: Analysis of 3746 records. Arch. Int. Med. *49:*345–359.
Viskum, K (1974) Peptic ulcer and pulmonary disease. Scand. J. Respir. Dis. *55:*284–290.
Wright, S, and Lewis, PA (1921) Factors in the resistance of guinea pigs to tuberculosis with especial regard to inbreeding and heredity. Am. Nat. *55:*20–50.
Wylie, CM (1981) The complex wane of peptic ulcer. J. Clin. Gastroenterol. *3:*327–332.
Yamaguchi, N, Agunod, M, Lopez, R, Luhby, AL, and Glass, GBJ (1968) Developmental patterns of hydrochloric acid, pepsin and intrinsic factor secretion in newborns and infants. In L Semb and J Myren (eds): The Physiology of Gastric Secretion. Baltimore: Williams and Wilkins, pp. 501–505.
Youmans, GP (1979) Tuberculosis. Philadelphia: Saunders.

AMERICAN JOURNAL OF PHYSICAL ANTHROPOLOGY 62:81–89 (1983)

The Human Histocompatibility System: Anthropological Considerations

GLENYS THOMSON
Genetics Department, University of California, Berkeley, California 94720

KEY WORDS HLA system, Disease association, Ankylosing spondylitis

ABSTRACT The human histocompatibility system (HLA) is a linked complex of genes on human chromosome 6. Many of the loci in this region are highly polymorphic. This endows the system with unique differentiating powers, both in terms of the population genetics of the reconstruction of evolutionary trees to assess biological divergences (or affinities) in human populations, and in terms of detecting the genetic component of the many diseases which show an association with certain variants (alleles) of the HLA system.

Different racial groups often exhibit different HLA disease associations; that is, a different allele is associated with the disease in different populations, although in some cases the same allele is associated with the disease in all populations. The classical example of this latter situation is the association of B27 with ankylosing spondylitis. This disease will be used as an example to illustrate how population observations allow inferences to be made regarding the evolutionary histories of the HLA-associated diseases, as well as the genetic mechanisms of the diseases.

The human histocompatibility system (HLA) was discovered as a blood-group-like system detected on the white cells of the blood. The impetus for the investigation of the HLA system was the need to match donors and recipients for the antigens that are important for tissue transplantation. Although the matching of donors and recipients for transplantation has turned out to be much more difficult and complex than was at one time hoped, out of this effort has grown our knowledge of the HLA system and its equivalents in other species.

The development of the HLA system has been greatly stimulated by a series of international collaborative workshops started in 1964. Each of these workshops has been of major importance in the development of the system, and its historical development is documented in the workshop proceedings (Histocompatibility Testing, 1965, 1967, 1970, 1972, 1975, 1977, 1980; see Van Rood, 1965; Curtoni et al., 1967; Terasaki, 1970; Dausset and Colombani, 1973; Kissmeyer-Nielsen, 1975; Bodmer et al., 1978a; Terasaki, 1980).

By a combination of family and somatic cell hybrid studies the HLA system has been mapped to the short arm of human chromosome 6. Chromosome 6 is one of the most extensively mapped human chromosomes. This is due in part to the large number of loci which have been mapped within the so-called HLA region. This region is depicted in Figure 1.

Within the HLA region are found the original serologically detected antigens of the HLA system. These antigens behave as if they were controlled by multiple alleles at three loci. These loci are defined as HLA-A, -B, and -C. The HLA-A and -B loci were discovered very early in the development of the system and are particularly well defined (see the workshop proceedings; Kissmeyer-Nielsen, 1975; Bodmer et al., 1978b). Another well-defined locus is HLA-D, which controls the mixed lymphocyte culture (MLC) response. What were initially called immune associated (Ia), and are now called DR (D related); serologically detected antigens are also found in the HLA region. It is believed that these antigens map to the HLA-D locus, or to a locus very close to D (Barnstable et al., 1979; Bodmer et al., 1978a). Throughout the follow-

Received August 5, 1982; accepted March 16, 1983.

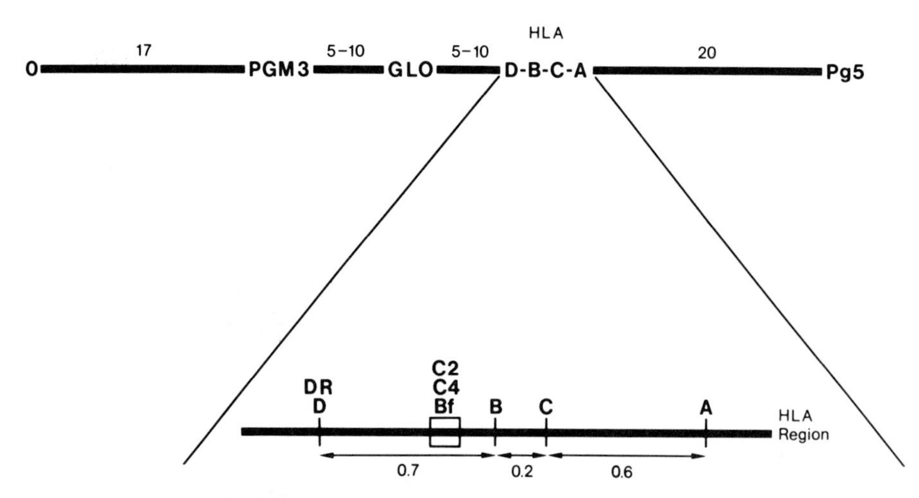

Fig. 1. The HLA Linkage Map and Chromosome 6. Based on Bodmer (1976) and Bodmer (1980). PGM3, phosphoglucomutase 3; GLO, glyoxylase; Pg5, pepsinogen 5; Bf, factor B of complement system; C2, second complement component; C4, fourth complement component. The positions of C2, C4, and Bf have not been finely mapped. (The Chido and Rodgers blood groups have been shown to be C4.) DR specificities are presumed to map mainly at, or close to, D. The distances are the approximate percentage recombination fractions.

ing discussion we will focus attention on these five loci, that is, the HLA-A, -B, -C, -D and DR loci.

On the basis of observed recombinants the map order of the loci is D, B, C, A, with DR mapping with, or close to, D (see Fig. 1). The recombination fraction between the D and B loci is .007, between the B and C loci .002, and between the C and A loci .006 (Bodmer, 1980). Calculations of Bodmer (1972) show that in humans there is enough DNA within a recombination fraction of .008 (the distance between the A and B loci) to code for about a 1,000 cistrons. So, even allowing for appreciable redundancy in the DNA it seems that there are a large number of genes still to be identified in this region.

At the population level the main feature of interest of the HLA region is the high level of polymorphism exhibited by many of the loci of the HLA system. At least 20 alleles (variants), also known as antigens, have been described for the A locus, 40 for the B locus, 8 for the C locus, 12 for the D locus, and 10 for the DR locus (Albert, 1980). We denote the alleles of these loci by a letter and number, the letter indicating the locus to which it is assigned—for example, A2 and A9 are alleles of the A locus, B7 is an allele at the B locus, etc. In no case is there a single very common allele; rather there are quite a few alleles with frequencies around 5–10%. All the alleles of these HLA loci have not as yet been determined. In the case of the A and B loci the frequencies of the blank alleles, representing as yet undetected determinants, are only about 2–4% in Caucasian populations, while the frequency of the blank at the other loci is much higher—for example it is around 50% for the C locus and 20–30% for the DR locus (Bodmer, 1980; Baur and Danilovs, 1980). The level of heterozygosity at each locus is very high, being around 90% for both the A and B loci (see the workshop proceedings; Bodmer, 1972, 1975, 1976; Bodmer and Thomson, 1977). Hedrick and Thomson (1983) have shown that the observed homozygosity for both the HLA-A and -B loci is always less than the expectation from neutrality and was statistically significantly less at the 0.05 level in 25 of 44 cases. They suggest that some form of symmetrical balancing selection is the explanation most consistent with the genetic variation at the A and B loci in the populations studied.

Theoretical explanations for a selective basis for the high level of polymorphism of the HLA loci have been put forward. It is argued that the function of the HLA loci may be such as to tend to give a temporary advantage to new variants (Bodmer, 1972). The line of reasoning for this argument involves the fact that a pathogen has to adapt to the antigenic or immune response characteristics of the host. A change in a host antigen mimicked by a pathogen, or used by it as a receptor, may necessitate a corresponding change in the pathogen for the pathogen to be successful in the altered host. So a host with an altered antigen will be at a selective advantage until the pathogen adapts to the new type, giving a form of frequency-dependent selection which could give rise to polymorphism. However, it must be stressed at this stage that these are theoretical speculations, and no direct evidence for such a mechanism exists.

There are a number of substantial differences in allele frequencies for the HLA loci in the major racial groups. These are documented in *Histocompatibility Testing 1980* (Terasaki, 1980) and *Histocompatibility Testing 1972* (Dausset and Colombani, 1973). For example, the antigen A1 occurs predominantly in Caucasoids where it has a frequency of around 15%, it is virtually absent in Japanese populations (0.5%) while its frequency of 3.3% in Black populations can probably be accounted for largely by Caucasoid admixture. In contrast, Aw30 is frequent in Black populations (15%), while it occurs with low frequencies in Caucasoid and Japanese populations, whereas Aw24 is much more frequent in Japanese populations than in Caucasoid or Black populations. The extreme variability of the loci in the HLA region and the fact that distinct patterns in allele frequencies exist between different racial groups, endows the system with unique differentiating powers in terms of the population genetics of the reconstruction of evolutionary trees to assess biological divergences (or affinities) in human populations (Schanfield, 1980).

However, while the HLA system is certainly one of the most useful genetic systems for anthropological studies, at present the scarcity of suitable antisera and hence the prohibitive cost of typing individuals, detracts from its utility. Hopefully in the near future this problem will be eased. At present most studies involve European, North American, and Japanese populations, and very few population studies on other groups have been published since the 1972 workshop, and even there the sample sizes for some studies were very small.

DISEASE STUDIES

A very exciting area of current HLA research comes from the demonstration that a number of alleles of the HLA system have been shown to be very significantly associated with a wide range of chronic diseases. By association here we mean that the population of individuals with the disease has a statistically significant increased frequency of a particular HLA antigen over a control population. The diseases reported to be associated with specific HLA antigens affect all organ systems. A list of these diseases, and the HLA antigen(s) associated with the disease, are given in Table 1. A common theme to these diseases is a possible autoimmune etiology. The existence of specific autoantibodies has been demonstrated in a number of diseases—for example, insulin-dependent diabetes mellitus (IDDM), Grave's disease, Addison's disease, ulcerative colitis, and others—and an autoimmune mechanism is suspected in a number of other HLA-associated diseases—for example, multiple sclerosis and rheumatoid arthritis.

The case of ankylosing spondylitis provides the most striking example of an HLA disease association. In Caucasian populations the frequency of the antigen B27 in patients is around 92% compared to 9% in controls (Kidd et al., 1977). (The alleles of the HLA loci are codominant, so to calculate the frequency of the antigen B27 in a particular group we include both individuals homozygous for B27 and those heterozygous for B27.) In the case of hemochromatosis, in Caucasian populations, the frequency of the antigen A3 is 75% amongst individuals with hemochromatosis, compared to 27% in a control population (Simon et al., 1977; Kidd, 1979). For multiple sclerosis the antigen DR2 shows the strongest association with the disease and is found in 67% of those with the disease, compared to 25% in a control group for a Caucasian population (Stewart et al., 1981). Results from the Eighth International Histocompatibility Workshop indicate that in Caucasian populations the antigen DR3 occurs in 59%, and the antigen DR4 in 77% of individuals with insulin-dependent diabetes mellitus, compared to frequencies of 21% and 22% in controls (Svejgaard et al., 1980). Many of the disease associations are not nearly as striking as these—for example, in acute lymphatic leukemia the antigen B8 occurs with a frequency of 29% in the individuals with the disease compared to 24% in the control group (McMichael and McDevitt, 1977). Nevertheless, these weaker associations are often consistently found as statistically significant associations.

In all populations that have been studied the antigen B27 has been found to be associated with ankylosing spondylitis (see Table 2). However, for some diseases, different racial groups exhibit different HLA disease associations. In the Japanese, where the frequency of the antigen DR3 is very low, the DR locus association with juvenile insulin-dependent diabetes is only with DR4, while for Caucasians the DR association is with DR3 and DR4 (Svejgaard et al., 1980). Similarly, in the Japanese, myasthenia gravis is associated with

TABLE 1. Diseases known to be associated with specific HLA antigens[1]

Disease		HLA-associated antigen
Rheumatology		
Ankylosing spondylitis	Caucasoids	A2, B27
	American Blacks	B27
	Persians	B27
	Japanese	B27
Reiter's disease	Caucasoids	B27
Yersinia arthritis	Caucasoids	B27
Salmonella arthritis	Caucasoids	B27
Psoriatic arthritis	Caucasoids	B13, B27, Bw38, B17
Frozen shoulder	Caucasoids	B27
Juvenile rheumatoid arthritis	Caucasoids	B27, B15
Acute anterior uveitis	Caucasoids	B27
Rheumatoid arthritis	Caucasoids	A2, B27, Dw4, DR4
Rheumatic heart disease	Caucasoids	A2
Neurology		
Multiple sclerosis	Caucasoids (except Italians)	A3, B7, Dw2, DR2
	Italians	A10, DR4, DR5
	Japanese	Bw22, DR5
Optic neuritis	Caucasoids	B7, Dw2
Myasthenia gravis	Caucasoids	A1, B8, Dw3, DR3
	Japanese	B12, DR4
Paralytic poliomyelitis	Caucasoids	Bw16
Schizophrenia	Caucasoids	A28, B5
Manic depressive disorder	Caucasoids	Bw16
Dermatology		
Psoriasis vulgaris	Caucasoids	B13, B17, B37
	Japanese	A1, B13
Psoriasis unspecified	Caucasoids	B13, B16, Bw17, Cw6
	Japanese	B37, Cw6
Pustular psoriasis	Caucasoids	B27
Dermatitis herpetiformes	Caucasoids	A1, A2, B8, Dw3
Pemphigus	Caucasoids	A10
	Japanese	A10
Bechet's disease	Caucasoids	B5
	Japanese	B5
Recurrent herpes labialis	Caucasoids	A1, B8
Alopecia areata		B12
Endocrinology		
Juvenile insulin-dependent diabetes	Caucasoids	B8, B15, B18, Cw3, Dw3, Dw4, DR3, DR4
	Japanese	Bw22, DR4
Thyrotoxicosis (Grave's disease)	Caucasoids	B8, Dw3
	Japanese	Bw35
Subacute thyroiditis	Caucasoids	Bw35, Dw1
Idiopathic Addison's disease	Caucasoids	B8, Dw3
Adrenocortical hyperfunction	Caucasoids	A1, A3
Gastroenterology		
Celiac disease (Gluten sensitive enteropathy)	Caucasoids	B8, Dw3
Ulcerative colitis	Japanese	B5
Pernicious anemia	Caucasoids	B7
Atrophic gastritis	Caucasoids	B7
Autoimmune chronic active hepatitis	Caucasoids	A1, B8, Dw3, DR3, DR4
Hepatitis B-associated chronic active hepatitis	Caucasoids	B18
Idiopathic hemochromatosis	Caucasoids	A3, B7, B14, Dw2, DR2
Allergology		
Dust allergy	Caucasoids	Aw33
Immunopathology		
Systemic lupus erythematosus	Caucasoids	B5, B8
Sicca syndrome	Caucasoids	B8, Dw3
Malignant diseases		
Retinoblastoma	Caucasoids	B12, Bw35
Hodgkin's disease	Caucasoids	A1, B5, B8, B18
Acute lymphatic leukaemia	Caucasoids	A2, B12

(Table 1. continued on next page)

TABLE 1. Diseases known to be associated with specific HLA antigens (continued)

Disease		HLA-associated antigen
Other diseases		
Congenital heart malformation	Caucasoids	A2
Polycystic kidney disease	Caucasoids	B5
Appendicitis acuta	Caucasoids	B12, B27
Asbestosis	Caucasoids	B27
Cryptogenic fibrosing alveolitis	Caucasoids	B12

[1]Adapted from Dausset and Svejgaard (1977) and Bodmer et al. (1978a). These data represent a summary of disease associations collected for the First International Symposium on HLA and Disease, Paris, 1976, and the Seventh International Histocompatibility Workshop and Conference, Oxford, 1977. Associations of HLA-C, -D, and -DR antigens have not been investigated for all diseases.

TABLE 2. The association of B27 with ankylosing spondylitis in various racial and ethnic groups[1]

		Patients		Controls				
Disease and racial group	HLA type	Sample size	% Pos-range	Sample size	% Pos-range	Relative risk	Significance	No. of studies
Ankylosing spondylitis								
Caucasian	B27	2,022	79–100	16,162	4–13	87.44	1.0×10^{-10}	29
Jewish	B27	38	79	456	3	109.50	4.5×10^{-30}	1
Japanese	B27	63	67–92	167	0–2	324.49	1.0×10^{-10}	2
Eastern Indians	B27	47	80–94	160	2–3	181.69	1.0×10^{-10}	2
Haida Indians	B27	17	100	222	51	34.38	1.7×10^{-5}	1
Bella Coola Indians	B27	3	100	129	26	20.16	1.9×10^{-2}	1
Pima Indians	B27	30	57	1,251	18	6.04	1.0×10^{-3}	1
Iranians	B27	25	92	400	3	349.59	4.4×10^{-28}	1
American Blacks	B27	23	48	60	2	36.49	8.5×10^{-7}	1
Venezualans	B27	19	58	303	3	41.94	1.3×10^{-10}	1

[1]Data from Ryder et al. (1979) and Calin (1976).

the antigen DR4 rather than with the antigen DR3, as in Caucasians.

These HLA disease associations have opened up new approaches which help to clarify the genetics and etiology of these diseases. (The diseases which show associations with the HLA system are ones which do not show simple Mendelian segregation, but nevertheless in many cases have an obvious inherited component, with incomplete penetrance due to the influence of environmental factors in predisposing individuals to disease.)

The study of different population groups can be very informative for our analyses. Again however, we are unfortunately hindered by lack of sufficient data on diverse ethnic groups.

It is sometimes observed that along with significant increases in the frequency of one or more antigens with a disease, there are significant decreases in frequency of some other antigens. For example, the antigens A5, B7, and DR2 are significantly decreased in Caucasoid individuals with insulin-dependent diabetes mellitus (Dausset and Svejgaard, 1977; Svejgaard et al., 1980), and the antigens A2 and B12 are decreased in Caucasoids with multiple sclerosis (Dausset and Svejgaard, 1977). When a decreased antigen frequency is observed the question arises as to whether these decreases are simply the inevitable result of the fact that if one or more alleles at a locus is increased in frequency then others must be decreased, or if the decreased frequency indicates that this particular allele provides some "protection" against the disease. Theoretical aspects of this question are currently being investigated (Thomson, Nicholas, O'Neill, Bodmer and Hedrick, work in progress).

MODES OF INHERITANCE OF THE HLA-ASSOCIATED DISEASE

The usual explanation that has been given to account for these HLA disease associations is that the association of a particular antigen(s) with a disease is the result of linkage disequilibrium (that is, nonrandom association) between the antigen(s) and the alleles at a nearby locus which confers susceptibility to disease. Under this assumption, two formal methods, the antigen genotype frequencies amongst diseased method and the affected sib pair method, have provided valuable insights into the mechanisms operating in disease susceptibility for some of the HLA-associated diseases (see Thomson, 1981, for a review of these methods).

Applying these methods to available data it is possible to determine that the HLA-linked disease-predisposing alleles in ankylosing spondylitis, multiple sclerosis, and dermatitis herpetiformis all follow basically a dominant mode of inheritance, whereas in hemochromatosis, insulin-dependent diabetes mellitus, and celiac disease the mode of inheritance of the HLA-linked disease predisposing allele is close to recessive. It is now clear that the original models investigated, which are based on a number of simplifying assumptions, are not always sufficient to explain the inheritance patterns of the HLA-associated disease. More refined and realistic models are currently being investigated. Nevertheless, the original models have given us a great deal of insight into the genetic mechanisms operating in disease susceptibility for some of the HLA associated diseases.

ANKYLOSING SPONDYLITIS

The question is often raised of whether it is actually the HLA antigens themselves which are directly involved in predisposing an individual to disease. This question is particularly relevant in the case of ankylosing spondylitis, where the association of the disease with B27 is very strong, and is found in all racial groups, and where there is evidence that the antigen B27 may cross-react with *Klebsiella pneumoniae* antigen in predisposing individuals to disease (Ebringer et al., 1978; Seager et al., 1979). Detailed data on the association of B27 with ankylosing spondylitis in various racial and ethnic groups are given in Table 2.

Ankylosing spondylitis is a disease which affects the sacroiliac joints and posterior intervertebral joints of the spine. In studies from the United States the estimated prevalance of the disease is given as about 0.1%, and the disease appears to be about six or seven times more frequent in males than females (Parker, 1980), although this may reflect ascertainment bias to some extent. If very mild disease is considered, the frequency of the disease is much higher, possibly as high as 2% in Caucasians (Parker, 1980).

The frequency of B27 and ankylosing spondylitis is reduced in American Blacks, compared to Caucasian populations. Also, the frequency of B27 amongst Blacks with ankylosing spondylitis is much less than that found in Caucasian populations (see Table 2). In contrast, in the Japanese, a mixed Israeli population, and in people indigenous to India, all of whom also have a reduced population frequency of B27, the frequency of B27 in ankylosing spondylitics is usually as high as that found in Caucasians.

In North American Indian groups that have been studied (Haida, Bella Coola, and Pima), the frequency of B27 is relatively high. In the Haida, where the frequency of B27 was 51% in the control population, a large proportion (10%) of the males were found to have ankylosing spondylitis, and of those males with ankylosing spondylitis, a total of 17, all had B27 (Gofton et al., 1975). The three Bella Coola ankylosing spondylitics were also B27 positive. In contrast, in the Pima population, where the control frequency of B27 was 18%, 20% of randomly selected individuals had ankylosing spondylitis, but of those with ankylosing spondylitis only 57% were B27 positive (Calin, 1982).

Thus, in the Caucasian groups, the Japanese and the Haida and Bella Coola Indians, individuals with ankylosing spondylitis exhibit a very high frequency of the antigen B27. We note that the Japanese control groups have a low frequency of B27 relative to the Caucasian groups, whereas the Haida and Bella Coola Indians have a high frequency. We contrast this to the associations of B27 with ankylosing spondylitis in the Pima Indian and American Black populations. In these two groups, the frequency of B27 amongst ankylosing spondylitics is relatively high (57% and 47%, respectively), but much lower than in the Caucasian groups, and whereas the Pima Indian control group has a high frequency of B27 relative to the Caucasian groups, the American Blacks have a low frequency.

The reduced frequency of B27 amongst American Blacks with ankylosing spondylitis has been taken as evidence to support the concept that it is not B27 itself which predisposes to ankylosing spondylitis, but a gene closely linked to the HLA-B locus, and in strong linkage disequilibrium with B27 (Sachs and Brewerton, 1978). This conclusion requires further investigation.

Note that the expected frequency of B27 amongst ankylosing spondylitics in a particular population group, if it is B27 itself which predisposes to disease, depends on two factors. These are the frequency of the predisposing antigen, B27 in this case, in the population, and the prevalance of environmental (or other genetic factors) which predispose to disease. These two factors will determine the incidence of the disease in the population and the frequency of B27 amongst individuals with the disease.

These points are illustrated by setting up a model where we assume that the antigen B27 itself predisposes to ankylosing spondylitis. Let p denote the frequency of the antigen B27 in the population, and put $q = 1 - p$, and we assume Hardy-Weinberg proportions for the B27 genotypes in the general population. We denote by f_2, f_1 and f_0 the respective probabilities that an individual who is homozygous B27, heterozygous B27, or who does not have the antigen B27, will contract the disease. Thus,

Genotype	B27/B27	B27/–	–/–
Frequency in the general population	p^2	^2pq	q^2
Probability of contracting disease	f_2	f_1	f_0

(1)

The frequency of the disease in the general population, denoted P(D), is thus

$$P(D) = f_2p^2 + 2f_1pq + f_0q^2 \quad (2)$$

and the frequency of B27 amongst individuals with the disease, denoted $f_D(B27+)$ is

$$f_D(B27+) = \frac{f_2p^2 + 2f_1pq}{P(D)} \quad (3)$$

It can be shown that if the probabilities f_2, f_1, and f_0 of contracting the disease, for a given genotype, remain constant from one population to another, and this could be taken to imply that the environmental predisposing factors, and other genetic factors which predispose to the disease remain constant, then the incidence of the disease (given by (2)) and the frequency of the antigen in the diseased group (given by (3)) will both always be less in populations with reduced frequencies of the antigen B27. Thus, the observation of reduced frequency of ankylosing spondylitis, and reduced frequency of B27 amongst ankylosing spondylitics, in American Blacks is in fact in agreement with the trend predicted by a model where B27 itself predisposes to disease. The data which are possibly in disagreement with these predictions are the Japanese data, the Israeli data, and the Pima Indian data.

The reduction in frequency of B27 amongst ankylosing spondylitics in a population where the frequency of B27 is reduced can be quite large. For illustration we consider a case where $f_2 = f_1 = 0.1$ and $f_0 = 0.001$. If we take the frequency of B27 in the general population as $p = 0.09$, to represent, say, a typical Caucasian population, then we would obtain, from (2) and (3), $P(D) = 0.018$ and $f_D(B27+) = 0.95$. When $p = 0.01$, representing, say, a U.S. Black population, then $P(D) = 0.003$ and $f_D(B27+) = 0.67$. (These figures are meant as illustration only to quantitate the magnitude of the trend predicted, and should not be compared directly with the figures given in Table 2.)

These predictions also relate in a straightforward manner to the so-called relative risk ascribed to a particular antigen with respect to the disease of interest. The relative risk (RR) is defined as

$$RR_A = \frac{P(D/A)}{P(D/\overline{A})} = \frac{P(A/D)\,P(\overline{A})}{P(\overline{A}/D)\,P(A)}, \quad (4)$$

where P(D/A) denotes the probability of disease given the presence of the antigen A, etc. In actual fact the relative risk is usually approximated by the odds ratio (OR) which is defined as:

$$OR_A = \frac{P(A/D)\,P(\overline{A}/\overline{D})}{P(\overline{A}/D)\,P(A/\overline{D})} \quad (5)$$

since in cases where the disease is rare these two quantities are approximately equal. The relative risks for B27 and ankylosing spondylitis in the various racial and ethnic groups are given in Table 2.

Under the present model, where we assume that B27 itself predisposes to disease, then

$$RR_{B27} = \frac{f_2p^2 + 2f_1pq}{f_0q^2}\,\frac{q^2}{p^2 + 2pq}. \quad (6)$$

Obviously, if B27 predisposes to disease in a dominant fashion, that is, $f_2 = f_1$ (and there is good evidence in the case of ankylosing spondylitis for a dominantly inherited predisposing disease allele, see Thomson, 1981, then

$$RR_{B27} = f_2/f_0. \quad (7)$$

This implies that if the parameter values f_2, f_1, and f_0 remain constant in all populations, then the relative risk for B27 should be the same in all populations, and independent of the population frequency of B27. One can see from Table 2 that the relative risks ascribed to B27 in the different racial and ethnic groups vary widely. Unfortunately, the standard errors on these estimates are so large that none of these values can be shown to be significantly different from each other. This results from the small sample sizes in the non-Caucasian studies.

However, it is clear from the above that to test the predictions of this model—that is, whether it is B27 itself which predisposes individuals to ankylosing spondylitis—one needs accurate estimates of the relative risks in the different population groups, the frequencies of B27 amongst the ankylosing spondylitics and the general population, and the incidences of the disease in the different populations. Combining all this information with expectations derived from equations (2), (3), and (7), one could then determine whether the data are compatible with the assumption that B27 itself predisposes to disease and that the environmental penetrance factors f_2, f_1, and f_0 remain constant in all populations. The preliminary data, outlined in Table 2, especially the Pima Indian data, indicate that if B27 itself predisposes to ankylosing spondylitis, then this latter assumption may not be correct.

It appears that present data on the frequencies of the disease, and the frequencies of B27 amongst ankylosing spondylitics and the general population, in different racial groups, and the resulting relative risk estimates, are inadequate at the moment to answer the question of whether or not it is B27 itself which predisposes individuals to ankylosing spondylitis. However, the predictions that can be obtained from equations (2), (3), and (7) indicate that with further data, a great deal of progress could be made in terms of understanding the mechanism of predisposition to the disease.

An argument which can be put forward to argue that it is not B27 itself which predisposes to ankylosing spondylitis is the following. While it is observed in Caucasian populations that more than 90% of individuals with ankylosing spondylitis have B27, the risk of an HLA-B27-positive individual developing ankylosing spondylitis is usually estimated to be about 5%, with a greater frequency in males than in females (Parker, 1980), while if very mild or atypical disease status are also considered then approximately 20% of B27-positive individuals of both sexes are affected (Parker, 1980; McDevitt, 1979). Even this latter figure is probably too low if B27 itself alone predisposes to the disease, given evidence that the penetrance of the disease, not including very mild or atypical cases, amongst B27-positive first-degree relatives of ankylosing spondylitics is of the order of 38% (Kidd et al., 1977). However, this does not completely rule out the possibility that it is the antigens themselves which are predisposing individuals to disease. The antigens as now defined may represent a heterogeneous class. The fact that the B27 cross-reacting antigens—for example, B7, B22, and B40—are commonly found in B27-negative ankylosing spondylitics would favor such a possibility (Calin, personal communication). Also, the data given above would be compatible with the notion that a second "disease" gene may be involved in predisposing individuals to disease. Such a two-locus disease model has been suggested for a number of other HLA-associated diseases, particularly insulin-dependent diabetes mellitus and celiac disease, as well as other diseases (see Thomson, 1981, for a discussion).

One of the strongest arguments in favor of the theory that B27 itself predisposes to ankylosing spondylitis is the fact that the B27 association with this disease is found in all racial groups. If one argues for an HLA-linked predisposing allele to ankylosing spondylitis which is distinct from B27, then it is necessary to hypothesize that the mutation to this disease-predisposing allele arose prior to the divergence of the major racial groups in humans, and has subsequently been maintained in all these populations. While not impossible, this is a somewhat unlikely event.

We see that evidence exists in favor of the hypothesis that it is B27 itself which predisposes to ankylosing spondylitis and also in favor of the hypothesis that disease predisposition is not due to B27 but to an HLA-linked disease predisposing gene. Further studies on different population groups, comparisons of family history of disease from B27-positive and B27-negative probands, as well as further immunological investigation of the B27 antigen and the B27 cross-reacting antigens will all aid in elucidation of this question. Such studies are currently in progress (Calin, personal communication).

DISCUSSION

The main point of the above analysis is to illustrate the inferences that can be made from racial and ethnic group comparisons in disease association studies, while at the same time emphasizing the necessity for explicitly detailing the assumptions inherent in a model, and the results which follow from such assumptions.

It is clear that the study of different racial groups can give us a great deal of insight into the genetic mechanism operating in disease susceptibility for the HLA-associated diseases. Such studies apply not only to the situation where we want to test if it is the antigen itself which predisposes to disease, but also to cases where disease predisposition is believed to be due to a disease susceptibility gene in the HLA region, and not to the antigens themselves. A combination of analyses conducted within a given population group and comparisons between population groups can provide valuable information.

This applies to many of the HLA-associated diseases. In particular, anthropological studies will be highly valuable for diseases such as multiple sclerosis where the frequency of the disease is known to vary widely between different population groups and to be related in a complex way to distance from the equator (Acheson, 1977). We are hampered by lack of data at the moment, but our understanding of the etiology of many of the HLA-associated diseases is increasing, and many exciting areas of research await us.

ACKNOWLEDGMENTS

This research was supported by NIH grant R01 HD 12731.

LITERATURE CITED

Acheson, ED (1977) Epidemiology of multiple sclerosis. Br. Med. Bull. *33:*9–14.

Albert, E (1980) Nomenclature for factors of the HLA system 1980. Tissue Antigens *16:*113–117.

Barnstable, CJ, Jones EA, and Bodmer, WF (1979) Genetic structure of major histocompatibility regions. Int. Rev. Biochem. *22*(Defense and Recognition IIA):151–255.

Baur, MP, and Danilovs, JA (1980) Population analysis of HLA-A, B, C, DR, and other genetic marchels. In P Terasaki (ed): Histocompatibility Testing 1980, pp. 955–1210. Univ. of Calif., Los Angeles.

Bodmer, WF (1972) Evolutionary significance of the HLA system. Nature *237:*139–145.

Bodmer, WF (1975) Evolution of HL-A and other major histocompatibility systems. Genetics *79:*293–304.

Bodmer, WF (1976) The HLA system. Excerpta Med. Int. Cong. Ser. *411:*295–307.

Bodmer, WF (1980) The HLA system and disease. J.R. Col. Physicians Lond. *14:*43–50.

Bodmer, WF, Batchelor, JR, Bodmer, JG, Festenstein, H, and Morris, PJ (eds) (1978a) Histocompatibility Testing 1977. Copenhagen: Munksgaard.

Bodmer, WF, Jones, EA, Barnstable, CJ, and Bodmer JG (1978b) Genetics of HLA: The major human histocompatibility system. Proc. R. Soc. Lond. [Biol.] *22:*93–116.

Bodmer, W. and Thomas, G (1977) Population genetics and evolution of the HLA system. In J Dausset and A Svejgaard (eds) HLA and Disease. Copenhagen, Munksgaard pp. 280–295.

Calin A (1982) High risk of sacroiliitis in HLA-B27 positive Pima Indian men. Arthritis Rheum. *25:*236–238.

Curtoni, ES, Mattiuz, M, and Tosi, R (eds) (1967) Histocompatibility Testing 1967. Copenhagen: Munksgaard.

Dausset, J, and Colombani, PJ (eds) (1973) Histocompatibility Testing 1972. Copenhagen: Munksgaard.

Dausset, J, and Svejgaard, A (eds) (1977) HLA and Disease. Copenhagen: Munksgaard.

Ebringer, RW, Cawdell, DR, Cowling, P, and Ebringer, A (1978) Sequential studies on ankylosing spondylitis: Association of Klebsiellia pneumoniae with active disease. Ann. Rheum. Dis. *37:*146–151.

Gofton, JP, Chalmers, A, Price, EE, and Reeve, CE (1975) HL-A27 and ankylosing spondylitis and B.C. Indians. J. Rheum. *2:*314–318.

Hedrick, P, and Thomson, G (1983) A Neutrality Test for HLA. Genetics (in press).

Kidd, KK (1979) Studies of the genetics of illnesses associated with HLA. In S Ferrone (ed): HLA Antigens in Clinical Medicine, pp 220–230, Garland Pub, New York.

Kidd, KK, Bernoco, D, Carbonara, AO, Daneo, V, Steiger, U, and Ceppellini, R (1977) Genetic Analysis of HLA-associated diseases: The "illness susceptible" gene frequency and sex ratio in ankyloing spondylitis. In J Dausset and A Svejgaard (eds): HLA and Disease. Copenhagen: Munksgaard. pp 72–80.

Kissmeyer-Nielsen, F (ed) (1975) Histocompatibility Testing 1975. Copenhagen: Munksgaard.

McDevitt, HO (1979) Genetic structure and functions of the major histocompatibility complex. In DJ McCarty (ed): Arthritis and Allied Conditions. Philadelphia: Lea and Febiger.

McMichael, M and McDevitt, H (1977) The association between the HLA system and disease. In: Progress in Medical Genetics. Series 2, Vol. 2, pp. 39–100, Saunders, Philadelphia.

Parker, CW (ed) (1980) Clinical Immunology. W.B. Saunders Co., Philadelphia, W.B. Saunders Co.

Ryder, LP, Andersen, E, and Svejgaard, A (eds) (1979) HLA and Disease Registry, Third Report. Copenhagen: Munksgaard.

Sachs, JA, and Brewerton, DA (1978) HLA, ankylosing spondylitis and rheumatoid arthritis. Br. Med. Bull. *34:*275–278.

Schanfield, MS (1980) The anthropological usefulness of highly polymorphic systems. In JH Mielke and MH Crawford (eds): Current Developments in Anthropological Genetics, Vol. I. Theory and Methods. New York: Plenum Press pp 65–82.

Seager, K, Bashir, HV, Geczy, AF, Edmonds, J and De Vere-Tyndall, A (1979) Evidence for a specific B27-associated cell surface marker on lymphocytes of patients with ankylosing spondylitis. Nature *277:*68–70.

Simon, M, Alexandre, J-L, Bourel, M, Le Marec, B, and Scordia, C (1977) Heredity of idiopathic haemochromatosis: A study of 106 families. Clin. Genet. *11:*327–341.

Stewart, GJ, McLeod, GJ, Basten, A, and Bashir, HV (1981) HLA family studies and multiple sclerosis: A common gene, dominantly expressed. Hum Immunol. *3:*13–29.

Svejgaard, A, Platz, P, and Ryder, LP (1980) Insulin-dependent diabetes mellitus. In P Terasaki (ed): Histocompatibility Testing 1980. pp. 638–656.

Terasaki, PI (ed)-(1970) Histocompatibility Testing 1970. Copenhagen: Munksgaard.

Terasaki, PI (ed) (1980) Histocompatibility Testing 1980. Copenhagen: Munksgaard.

Thomson, G (1981) A review of theoretical aspects of HLA and disease associations. Theor. Popul. Biol. *20:*168–208.

Van Rood, J (1965) Histocompatibility Testing 1965. Copenhagen: Munksgaard.

AMERICAN JOURNAL OF PHYSICAL ANTHROPOLOGY 62:91–105 (1983)

Genetic and Sociocultural Components of High Blood Pressure

R.H. WARD
Department of Medical Genetics, University of British Columbia, Vancouver, British Columbia V6T 1W5

KEY WORDS Blood pressure, Genetic Epidemiology, Migrants

ABSTRACT The cardiovascular diseases exert widely differing contributions to the total burden of mortality and morbidity in extant human populations. To a large extent these differences are a reflection of the variable distribution of specific antecedent risk factors. For one such risk factor, blood pressure, there is considerable variability in its distribution between different ethnic groups, especially between traditional and nontraditional societies. Intensive epidemiological studies in Western societies, together with a number of cross-cultural comparisons, suggest that the major determinants of high blood pressure are likely to be a constellation of sociocultural factors, with genetic determination being limited to the interaction between genotype and environment.

Studies of populations in sociocultural transition offer an unique opportunity to identify the relative influence of specific sociocultural factors on the rate of change of blood pressure. In addition, when the study of such populations is placed in a quasi-experimental context, genetic-environmental interactions may also be detected. This strategy is illustrated by a study of the changing blood pressure distribution in Tokelauan migrants.

Such an approach requires the initial definition of a *response variable* which measures change in blood pressure as a consequence of migration. The response variable, which identifies the relative influence of concomitants such as weight, age, and obesity, can then be subjected to genetic analysis. In the Tokelau case, blood pressure response tends to be positive in migrants but negative in nonmigrants. Further statistical analysis indicates that there is a small proportion of high responders in both populations and that these cluster in families in the migrant population. However, estimates of the transmission parameter suggest that sociocultural transmission, rather than Mendelian segregation, is responsible. To date there is little evidence that genetic-environmental interactions have had any impact on the development of hypertension in this migrant population.

Nowadays it hardly needs to be emphasized that high blood pressure, or hypertension, is a significant risk factor for the variety of cardiovascular diseases which impose a significant burden of ill health for many populations around the world. While the particular manifestations of cardiovascular disease may vary from one society to another, the overall impact is highly significant in terms of increased mortality and increased morbidity. It is also recognised that the different manifestations of cardiovascular disease in different societies are undoubtedly a reflection of differing distributions of risk factors, such as blood pressure. These risk factors are in turn influenced by the variability in "life styles" and genetic susceptibility that characterise different populations.

While extreme hypertension has long been recognised as an important clinical problem, the identification of elevated blood pressure as a primary risk factor for cardiovascular disease is much more recent. This delayed recognition

Received August 5, 1982; accepted March 16, 1983.

is partly attributable to the fact that as long as infectious disease formed the major component of mortality, cardiovascular disease and its associated risk factors received relatively little attention. It was also partly due to the mistaken assumption that the increasing occurrence of hypertension and cardiovascular disease in older people was merely a natural consequence of ageing. Such a view is now known to be erroneous and comparative studies of different societies suggest that a rise in blood pressure with age is neither inconsequential, nor inevitable.

The relationship between high blood pressure and heart disease started to accumulate nearly forty years ago. The studies of Davis and Klainer (1940) suggested a close correlation between levels of blood pressure and pathological evidence of coronary heart disease. Later, Bechgaard (1946) showed that 45% of deaths in hypertensive patients were attributable to coronary heart disease—a finding substantiated by many later studies. Subsequent epidemiological studies have consistently shown that hypertension is associated with a marked and significant increase in the risk of cardiovascular disease in both men and women. This has been most clearly demonstrated in longitudinal studies such as the Framingham cohort study (Kannel et al., 1969), and large surveys such as the Chicago studies (Stamler, 1980). Comparative investigations across different sociocultural environments indicate that the form of cardiovascular disease may vary widely (Dreyfuss et al., 1961; Keys, 1970), and that the interaction of other risk factors, such as hypercholesteremia, with hypertension also need to be considered (Stamler et al., 1975).

HIGH BLOOD PRESSURE AND SOCIOCULTURAL FACTORS

The realisation that blood pressure levels may differ widely from one society to another goes back somewhat more than 60 years. Initial observations by Donnison (1929) and Vint (1937) indicated that unlike the situation in white populations, essential hypertension was extremely rare in certain black populations in Africa. The supposition that this finding was due to a genetic difference between blacks and whites was quickly dashed by the observation that the prevalence of hypertension in U.S. blacks was approximately double that observed in U.S. whites (Keith et al., 1939). Sociocultural factors rather than genetic background then seemed a more likely explanation for these marked differences in the distribution of blood pressure. This general conclusion has been substantiated by more comprehensive surveys in Africa which indicate that there is considerable variation in the distribution of blood pressure throughout the continent. In general, the prevalence of hypertension appears to be lowest in populations which still retain some traditional aspects of a tribal society, and is higher in populations which have adopted a more western and urbanised mode of life (Schrire, 1958; Abrahams et al., 1960; Gampel et al., 1962; Scotch, 1963; Shaper, 1967; Akinkugbe, 1969 1976; Akinkugbe and Ojo, 1969; Johnson, 1971; Levitt et al., 1974; Vaughan and Miall, 1979).

During the past 30 years, other studies in a variety of non-African societies have corroborated the general conclusion that hypertension is rare in traditional societies, though exceptions do occur (Neilson and Williams, 1978; Marmot, 1979). The evidence from a number of studies in the South Pacific is particularly striking. Murphy (1955) first drew attention to the fact that traditional societies of the tropical Pacific were characterized by low levels of blood pressure and little rise of blood pressure with increasing age. These preliminary observations have now been substantiated for all the major ethnic groups in the South Pacific which still retain some adherence to a traditional, pre-European, lifestyle: Polynesian (Maddocks, 1961; Prior, 1970; Prior et al., 1966), Melanesian (Page et al., 1974), and Papuan (Whyte, 1958; Sinnet and Whyte, 1973). By contrast, those Pacific societies that have substantially adopted a Western way of life exhibit elevated levels of blood pressure with an increase of blood pressure with age (Prior et al., 1966; Prior, 1970).

This relationship between the overall level of blood pressure and the rate at which blood pressure rises with age was first systematically classified by Epstein and Eckoff (1967). In their classification system, the majority of populations which exhibit low mean blood pressure have "zero" slope with respect to the relationship between blood pressure and age. As mean blood pressure levels increase, the rate of blood pressure increase with age also rises. Epstein and Eckoff's classification clearly shows that the majority of traditional societies are characterised by an insignificant increase of blood pressure with age, in addition to having low mean values of blood pressure.

Despite this general finding, the underlying reason for the lack of high blood pressure in nonwesternised societies is still not clear. While a large number of factors may be implicated, among the most important to be enumerated are salt use, degree of obesity, amount of physical activity, and psychological stress. Of these factors, low salt intake appears to play the dominant role for certain unacculturated societies. This is most dramatically demonstrated by the Yanomama (Oliver et al., 1975; Oliver, 1980) but low salt use is also implicated as the cause of low blood pressure levels in some other traditional societies (Page, 1980; Prior et al., 1968; Prior and Stanhope, 1980). However, even in the extreme case of the Yanomama, the role of the other three factors noted above cannot be discounted.

One tenable generalisation is that although risk factors for hypertension cut across cultural patterns, geographic regions, and stages of political development (Marmot, 1979), traditional societies are characterized by a lifestyle and sociocultural environment which results in a minimal influence of the above four major determinants of elevated blood pressure. Conversely, many westernized societies are characterized by lifestyles and a sociocultural environment which lead to a high frequency of hypertension. Curiously, societies in transition from traditional cultural patterns to urbanised Western culture often display a more extreme distribution of the determinants of hypertension than the Western societies that they are beginning to emulate. As a consequence, the epidemiology of hypertension and other cardiovascular risk factors is likely to be particularly important for societies in sociocultural transition.

IS THERE A GENETIC ROLE IN HIGH BLOOD PRESSURE?

Despite the overwhelming evidence of the importance of sociocultural factors in influencing the distribution of high blood pressure, the role of genetic factors cannot be discounted. For over 2 centuries, familial aggregation of cardiovascular disease has been recognised and, more recently, the familial aggregation of blood pressure has also been noted. Forty years ago, the familial clustering of hypertension was initially interpreted as Mendelian segregation of a major locus. However, once it was appreciated that the hypertensive patient actually originated from the tail of a continuous distribution of blood pressure, a polygenic model seemed a more likely explanation for the observed familial aggregation (Pickering, 1967). This is now the majority viewpoint and "hypertension" is universally recognised to be an arbitrary demarcation in the continuum of blood pressures. Even so, modern techniques, such as the identification of differences in cation flux between hypertensives and normotensives (Garay et al., 1980; Canessa et al., 1980), may yet reveal the existence of real discontinuities in the distribution of blood pressure and, once more, focus attention on the presumptive role of major genes.

Another line of evidence militating against single genes as the intrinsic factors underlying high blood pressure comes from the cross-cultural studies referred to above. There is a uniform association of low blood pressure levels with traditional tribal societies. If the four key factors characterising traditional societies (low salt use, leaness, high levels of physical activity, and low levels of psychosocial stress) are retained, there is relatively little variability in blood pressure levels across a wide variety of genetically distinct populations. This is in marked contrast to the extensive genetic variability between tribal populations, as indicated by the distribution of genetic polymorphisms. The relative uniformity in blood pressure levels also contrasts with the interpopulation variability in morphological characters, including those known to have high heritability, such as height and other anthropometric measures. However, when tribal societies become acculturated and deviate from the sociocultural milieu that characterises nonwesternised, nonurbanised societies mean blood pressure rises and interpopulation variability in blood pressure becomes more apparent. This suggests that on a global level, the major differences in blood pressure levels between populations is due to sociocultural differences, rather than genetic heterogeneity.

However, when the variability of blood pressure *within* populations is studied, a large number of investigations indicate that a fairly constant proportion of this variability (20–40%) is attributable to familial aggregation, which has been interpreted as the action of polygenic factors (Feinleib et al., 1980). Hence, within populations, variability in blood pressure may still be mediated by genetic factors, even if the more dramatic variability between populations is largely attributable to environmental factors.

Another kind of gene action which cannot be discounted is the influence of genetic-environmental interactions which may also constitute an important source of the observed variability in blood pressure distributions within populations. Although the major differences between societies appear to be a consequence of rather rapid environmental change, the distribution of hypertension within such societies is far from uniform. There is some precedent from genetic studies of other diseases for believing that many of the individuals who develop hypertension in "high-risk" societies may be genetically susceptible to a particular environmental risk factor. An evolutionary perspective also suggests that genetic-environmental interactions may play a role in influencing the present day distribution of high blood pressure and possibly other cardiovascular risk factors.

For most of man's history, high blood pressure and other cardiovascular risk factors made only an insignificant contribution to the overall burden of mortality. Only during the last 100 generations were these complex multifactorial traits likely to have had a significant impact. Thus, any specific genetic influences on the distribution of cardiovascular risk factors probably arose indirectly. Genetic traits that were not harmful in the original type of environment characteristic of tribal societies may have accumulated by the process of random drift and then exerted their detrimental effect when the environment changed. Alternatively, as first suggested by Neel (1962), selective pressures in the more precarious environments of our ancestors may have favoured the spread of specific metabolic processes which subsequently turned out to be deleterious in our present-day environments, which are characterised by an abundance of food, salt, and stress. Such an evolutionary viewpoint tends to support the hypothesis that interactions between susceptible genotypes and deleterious environments are important in the development of hypertension.

To summarize: Our own species exhibits a peculiar tendency for the development of high blood pressure, with attendant clinical complications. The extensive distribution of hypertension, and other cardiovascular risk factors, is more likely to be a consequence of cultural evolution than a product of our biological evolution. Further, the relatively recent increase in the frequency of hypertensive individuals is likely to be partly attributable to genetic-environmental interactions, especially with regard to their occurrence within populations. Hence, in attempting to identify the determinants of high blood pressure in societies which are undergoing rapid cultural transition, the potential impact of interactions between environment and genotype cannot be neglected.

IDENTIFYING GENOTYPIC-ENVIRONMENTAL INTERACTIONS: THE QUASI-EXPERIMENTAL APPROACH

One of the most severe limitations of genetic epidemiology is the difficulty of inferring genetic causation from observational studies (Kempthorne, 1978). This problem becomes particularly acute if genotype-environmental factors are thought to influence the observed distribution of the trait, as is likely for blood pressure. In the absence of a truly experimental approach, a "quasi-experimental" study design will be required if causation involving interaction is to be inferred (Ward et al., 1979). Elsewhere we have argued that a "quasi-experimental" study design can result from an appropriately conducted investigation of migrant isolates (Ward and Prior, 1980).

In the context of studying the determinants of high blood pressure, migrating isolates are particularly favorable since they normally comprise groups who are eschewing a traditional milieu for a more urbanized and westernized environment (Cassel, 1974). In so doing, they usually experience rapid sociocultural transition and tend to be characterized by a relatively rapid increase in the levels of high blood pressure. Thus, this type of study design is particularly suited for the study of the interaction between genotype and environment since a marked rise in the frequency of individuals with high blood pressure is a predictable response to the migration, and its accompanying sociocultural change. Since sociocultural transition may occur in situ, this approach does not necessarily require studying a community that is undergoing geographic dislocation. However, our own studies have concentrated on a small population, Tokelau, that has undergone migration from its traditional atoll homeland to New Zealand.

The Tokelau Island Migrant Study is especially suited for the identification of genetic-environmental determinants of blood pressure rise. It is a longitudinal study, spanning nearly 20 years, and the migration is from a known "low-risk" environment to a presumed "high-risk" environment. The study has the advantage that the population was extensively studied before the bulk of migration occurred. This allows a ready identification of an appropriate measure of response (see below). While the epidemiological results of this interesting study design have been published elsewhere (Prior et al., 1974, 1977) some pertinent results are relevant here: Studies of the premigrant population suggest that migrants were essentially a random sample of the original population with respect to the distribution of cardiovascular risk factors. In particular, before migration occurred the potential migrants and nonmigrants exhibited no significant differences in their blood pressure (Stanhope and Prior, 1976; Prior et al., 1977). Subsequent follow-up of migrants and nonmigrants, in their respective environments, indicates that migrant children and adults display a marked increase in the frequency of hypertension and other cardiovascular risk factors compared to nonmigrants (Beaglehole et al, 1977a,b, 1978, 1979; Stanhope and Prior., 1976; Stanhope et al., 1981; Ward et al., 1980a,b).

Identifying the underlying causes of this divergence in the distribution of blood pressure between migrants and nonmigrants has been more difficult. One important factor appears to be the increased obesity amongst migrants, though sociocultural stress factors may also play a small role (Beaglehole et al., 1978). Marked familial aggregation of blood pressure exists in both environments (Beaglehole et al., 1975; Ward et al., 1979 1980a,b). One interesting feature is that the degree of familial aggregation appears different in the two environments, with the transmission of sociocultural factors being markedly different in the migrant population compared with the nonmigrants (Ward et al., 1980a,b; Ward and Prior, 1980).

An original objective of this study was to determine whether genetic factors influence the marked rise in blood pressure experienced by the migrants. In this context, some effort must be made to identify the role played by genetic-environmental interactions. We are attempting to do this by investigating the distribution of blood pressure *response* in both migrants and nonmigrants. Our definition of blood pressure response is briefly outlined below.

DEFINITION OF RESPONSE VARIABLE

As noted previously, a "quasi-experimental" study design can only be achieved if blood pressure is measured on sets of related individuals distributed across different environments, where the related individuals have a genetic covariance structure defined in terms of the coefficient of kinship (Ward et al., 1979). In addition, if genotype-environmental interactions are to be identified there must be a longitudinal component to the study design, so that blood pressure change over time in each of the two environments can be evaluated. Ideally, the study design would include a comprehensive survey of all individuals in the old environment, before any migration took place. Also, the ideal migration event would be equivalent to a random sampling of genotypes living in the old environment. Subsequently, surveys would be made of the two random samples of related genotypes, each in distinct and changing environments. If these sampling conditions are met, then the covariance matrix of the coefficient of kinship represents a reasonable surrogate measure of genotypic identity of the related individuals, within and between the environments. Blood pressure changes in the different environments can be evaluated in terms of the response of specific, though arbitrary, genotypes to measured environmental stimuli. Such a study design may thus lead to estimates of genotype-environmental interaction, along with other more familiar variance components, as these quantities are classically defined (Haldane, 1946; Turner and Young, 1969).

Under these situations, the measure of response for every individual could be explicitly written in terms of the vector of blood pressure measurements made over time, referred back to the baseline measurement made before migration took place. Factors of secular change, such as the ageing that would occur, would be taken account of by appropriately standardizing the measured change in blood pressure for changes in age and other relevant concomitant variables.

Unfortunately, such ideal situations rarely, if ever, obtain, and the Tokelau study is no exception. In the less-perfect world of actual surveys, we are usually faced with the twin problems that not all individuals were measured before the migration event took place and that not all individuals are necessarily measured at each longitudinal point in time. In order to deal with these problems we have developed a statistical model to use as an approximation to the ideal situation described above (Altman et al., Ward et al.,).

A useful response variable will utilize information from the premigrant survey in such a way that the subsequent distribution of blood pressure can be partitioned into two specific components. One component will represent aspects of secular change (such as ageing), which are of essentially little interest to the geneticist, while the second, representing components of interaction in the new environment, will be of considerably genetic interest. The latter component can be further partitioned as follows: First, the changes in blood pressure may arise as a consequence of an altered relationship between a concomitant variable (or a set of concomitant variables) and blood pressure. Thus, if the relationship between weight and blood pressure changes after migration this can be viewed as an interaction component. Second, blood pressure in the new environment may be influenced by factors which were not present in the old environment. As shown below, both types of response can be derived in such a way that the influence of secular trend is eliminated.

We achieve these partitions by first specifying the routinely used epidemiological model, such that in the original premigrant environment blood pressure is defined in terms of a linear function of a set of measured concomitant variables, such as age, sex, atoll of origin, weight, and so forth, plus an unmeasured error component. The relationship between blood pressure and the concomitants can then be formally represented by the following equation:

$$BP_1 = f(\text{sex, age, atoll, weight, etc.}) + \text{error}$$

or, for the jth individual

$$BP_{1j} = \beta_1^T(X_{1j}) + e_{1j} \tag{1}$$

where X_{1j} refers to the vector of concomitants for the jth individual before migration, β_1^T is the transposed vector of regression coefficients for these concomitants, and e_{1j} is the "error" term.

In the new environment, under the generalized model of interaction, the distribution of blood pressure will now be influenced by two distinct vectors: First, there will continue to be a functional relationship between blood pressure and the original set of concomitants that were measured in the old environment, but the functional relationship will not necessarily be the same. In addition, there may be a functional relationship between blood pressure and a new set of concomitants which were not present in the old environment. As before, there will also be an error term. Hence, the formal specification of blood pressure in the new environment may now be written as follows:

$$BP_2 = h(\text{sex, age, atoll, weight, etc.}) + y(\text{new concomitants}) + \text{error}$$

However, assuming linearity, the new functional relationship between the set of concomitants that were measured in the old environment and blood pressure can be split into two parts. The first part represents the original relationship between the concomitants while the second part represents a new functional relationship between concomitants and blood pressure. When summed, these two functions determine the overall relationship. Using this partition, the distribution of blood pressure in the new environment can now be represented as follows:

$$\begin{aligned} BP_{2j} &= h(X_{2j}) + y(U_j) + \text{error} \\ &= f(X_{2j}) + g(X_{2j}) + y(U_j) + e_{2j} \end{aligned}$$

or

$$BP_{2j} = \beta_1^T(X_{2j}) + \beta_2^T(X_{2j}) + \alpha^T(U_j) + e_{2j} \tag{2}$$

From this last equation, it is clear that one way of defining response in the new environment is to define an initial functional relationship between blood pressure and a set of concomitants in the original environment, and use this to evaluate the deviations of observed blood pressure in the new environment from the predicted value. This deviation, tantamount to a derived residual, will be termed the "full response" as it incorporates both the changed functional relationships between the concomitants of the old environment and blood pressure, plus the influence of the new variables that were not present in the old environment. Formally, we define the full response as follows:

$$\begin{aligned} R_{F_j} &= BP_{2j} - f(X_{2j}) \\ &= BP_{2j} - \beta_1^T(X_{2j}) \\ &= \beta_2^T(X_{2j}) + \alpha^T(U_j) + e_{2j} \end{aligned} \tag{3}$$

Since the set of concomitants is available for each individual, this full response can be standardized for the influence of the concomitants entering into the original equation. This standardization is equivalent to removing the influence of the changed relationship between the concomitants and blood pressure. Thus the standardized response for each individual will only incorporate the influence of new factors which were unmeasured in the old environment. This is given by the following equation:

$$\begin{aligned} R_{S_j} &= R_{F_j} - \beta_2^T(X_{2j}) \\ &= \alpha^T(U_j) + e_{2j} \end{aligned} \tag{4}$$

An alternative way of defining the response variables is in terms of the actual change in blood pressure. Given the above definitions, this can be readily done as follows. The change in blood pressure over time is merely the difference between the two functional forms of blood pressure given by equations (1) and (2) above. As shown below, this leads to the following relationship: The change in blood pressure over time is the sum of the full response variable, as defined above, plus the functional form of the actual change in concomitants that occur during the two time periods.

$$\begin{aligned}\Delta_{BP} &= BP_2 - BP_1 \\ &= (\beta_2^T(X_2) + \beta_2^T(X_2) + \alpha^T(U)) - (\beta_1^T(X_1)) \\ &= \beta_1^T(X_2 - X_1) + \beta_2^T(X_2) + \alpha^T(U) \\ &= f(\Delta_X) + R_F \qquad (5)\end{aligned}$$

Alternatively, the full response can be considered as the change in blood pressure minus the functional form of the change in concomitants.

$$R_F = \Delta_{BP} - f(\Delta_X) \qquad (6)$$

Since the concomitants involve factors such as age, weight, and so forth, any change in blood pressure due to these factors will be of no genetic interest. Hence, the response variable will measure components of the change in blood pressure that are due to an interaction of concomitants within the new environment, plus the addition of new factors, rather than merely incorporate the inevitable changes due to secular trends. Thus there is some justification in assuming that the distribution of this particular measure after migration has occurred may be of considerable genetic interest. Below, we apply this model to data derived from the Tokelau migrant project, which is particularly suited to this type of analysis.

DEFINING RESPONSE IN TOKELAUAN MIGRANTS

Before defining the response variable, as outlined above, we wished to ascertain whether the variables that had previously been identified as important concomitants for blood pressure were relevant for blood pressure change. We did this by analyzing the change from the blood pressure values initially observed in 1968/71, for 543 nonmigrant adults subsequently measured in 1976, and 291 migrant adults measured in New Zealand in 1975/77. The difference between the second blood pressure measurement and the first was then regressed against the initial blood pressure value; subsequent changes in age, weight, height, percent fat, body mass and serum cholesterol; plus sex and atoll of origin. Interactions between sex, atoll of origin, and the various concomitants were also examined.

The results of the step-wise regression analysis of the change in systolic blood pressure for migrants and nonmigrants are displayed in Table 1. Inspection of the R^2 values indicates that the factors associated with blood pressure changes are different in the migrant and nonmigrant populations, and vary by atoll and sex. For example, previous systolic pressure is an important determinant of subsequent blood pressure change for nonmigrants (especially males from Fakaofo and Nukunonu) but is less important for migrants. Change of blood pressure in migrant males from Fakaofo and Atafu

TABLE 1. Factors associated with change in systolic blood pressure for Tokelauan migrants and nonmigrants, stratified by atoll of origin. Values in table represent the R^2 values associated with stepwise multiple regression of change in systolic blood pressure on the concomitant variables, with the variables entering the equation in the order listed

	Atoll of Origin		Previous systolic blood pressure	Change in: % Fat	Body Mass	Cholesterol	Age	Age²	Total
Males									
	Fakaofo	Migrants	0.0	21.6*	16.9*	4.5	12.4	9.5	64.9**
		Nonmigrants	33.7*	0.7	5.8*	0.8	0.0	10.0*	51.0**
	Nukunonu	Migrants	17.9	0.0	0.0	7.9	8.4	4.9	39.1
		Nonmigrants	26.8*	2.5	4.8	2.1	0.6	0.4	37.2**
	Atafu	Migrants	0.0	32.1*	3.2	0.8	1.5	0.4	38.0
		Nonmigrants	2.9	0.7	1.3	0.8	0.3	1.7	7.7
Females									
	Fakaofo	Migrants	0.2	0.8	15.6**	19.7**	17.3**	1.5	55.2**
		Nonmigrants	7.5**	0.3	12.2**	0.0	10.5**	0.0	30.5**
	Nukunonu	Migrants	18.2*	2.8	5.1	3.9	18.4*	3.3	51.7*
		Nonmigrants	6.0*	1.4	1.8	0.2	0.5	0.2	10.1
	Atafu	Migrants	3.2	5.0	21.3**	14.2*	5.5	4.1	53.3**
		Nonmigrants	0.4	0.1	8.0*	0.1	4.6	0.4	13.6

*p < .05.
**p < .01.

is related to changes in percent fat, while for migrant females, changes in body mass and cholesterol values are more important. In general, changes in concomitant values, such as weight and obesity, appear to be more important determinants of blood pressure change in migrants than in nonmigrants.

These results in Table 1 indicate that those concomitants associated with the variability of systolic blood pressure in the premigrant population are associated with the variability in blood pressure change subsequent to migration. A similar result obtains for the factors associated with changes in diastolic blood pressure, with the pattern of association varying by sex, atoll, and migrant status (Altman et al., Ward et al.,). This information is used to ensure that the response variable is defined in terms of those concomitants, which are relevant to blood pressure change. The results of Table 1 also indicate that the response equation has to be defined separately by atoll of origin and sex.

The standardised premigrant relationship between blood pressure and concomitants (the function, f(x), above) was defined using data from the 1968/71 survey, employing the age cut-offs that define adult blood pressure (Ward et al., 1979). This analysis was carried out for 984 individuals for diastolic blood pressure and 1,069 for systolic blood pressure. We used the full set of concomitants identified previously (Ward et al., 1979, 1980a,b), as well as the full set of interactions with age, sex, and atoll of origin.

The full regression equation accounted for essentially half the variability in initial systolic blood pressure (49.5%) but only a third of the variability in diastolic blood pressure (33.0%). Since the full model, with 53 separate coefficients, was rather cumbersome, we defined a reduced model incorporating only six terms. This accounted for 44.3% and 27.1% of the variability in premigrant systolic pressure and diastolic pressure, respectively. The important concomitants in this equation were body mass and percent fat, plus age, age^2 and age^3. While there were differences in the relationship between systolic blood pressure and age for males and females, the slope of the relationship between blood pressure and concomitants was the same for all atolls. The mean difference between atolls spanned 4 mm of mercury from Atafu (the lightest) to Nukunonu (the heaviest). For diastolic blood pressure there were no mean differences between atolls but there were differences in slope between males and females. In addition, there were small interaction terms between atoll and percent fat and age for males.

In calculating the blood pressure response, the observed blood pressure was used for the following groups of individuals: For nonmigrants (surveyed in Tokelau in 1976) there were 445 females 14 and older, 350 males of that age (diastolic response), but only 278 males 17 and older (systolic response); for the migrant population (surveyed in New Zealand between 1975 and 1977), there were 524 females 14 and older, 636 men of that age (diastolic response), and 540 males 17 and older (systolic response). For each of these individuals, the predicted blood pressure was calculated according to the equation referred to above, and subtracted from the observed value. This gave the full response variable for that individual.

The distribution of full blood pressure response in migrants and nonmigrants is given in Table 2, stratified by atoll of origin. Inspection of this table indicates that the value of the full response is quite different in the migrants and nonmigrants (as expected), but also varied by atoll and sex. It is noteworthy that the mean response values are low for migrant males and negative for migrant females and all nonmigrants. This indicates that relative to the changes that have occurred in concomitants subsequent to migration, the change has been surprisingly modest. Thus the blood pressure values after migration tend to be *lower* than the values expected on the basis of observed changes in concomitants.

Relative to the atoll of origin, migrants tend to have higher blood pressures than nonmigrants, with correspondingly higher response values. For males, the difference between migrants and nonmigrants is between 5 and 10 mm mercury for both systolic blood pressure and diastolic blood pressure. This is approximately the age-standardized difference between migrants and nonmigrants for systolic blood pressure, but somewhat higher than that observed between migrants and nonmigrants for diastolic blood pressure. For females, the difference between response scores, taking atoll of origin into account, is somewhat less, being approximately 1–4 mm for systolic blood pressure and 1.6–5 mm for diastolic blood pressure. However, these differences in response values are also similar to the age-standardized differences between migrant and nonmigrant females. Since the magnitude of the difference in response scores between migrants and nonmigrants is similar to the observed difference

TABLE 2. Distribution of full blood pressure response in Tokelauan migrants and nonmigrants, by atoll of origin

		Migrants			Nonmigrants		
		Atafu	Fakaofo	Nukunonu	Atafu	Fakaofo	Nukunonu
Males							
	Systolic						
	Mean	0.55	4.53	0.54	−4.96	−5.60	−8.99
	S.D.	14.9	14.9	15.2	14.8	13.9	16.6
	Diastolic						
	Mean	3.96	3.79	6.90	−5.88	−3.04	−3.25
	S.D.	11.1	10.2	10.0	9.1	9.6	9.6
Females							
	Systolic						
	Mean	−5.12	−2.60	−4.05	−6.07	−5.61	−6.53
	S.D.	14.9	18.1	13.6	15.8	14.4	15.5
	Diastolic						
	Mean	−2.47	−2.78	0.69	−6.79	−4.36	−4.57
	S.D.	9.9	10.8	9.4	10.2	11.2	9.6

in blood pressure values, the factors underlying blood pressure change after migration must involve a changed relationship between concomitants and blood pressure, as suggested by Ward et al. (1980b).

The relatively large standard deviations indicate a great deal of dispersion of response score about the mean values. To determine whether the variability in response score was a function of the actual change in blood pressure, we compared response scores and blood pressure change for migrants and nonmigrants with blood pressure data from the original premigrant survey. Response scores were stratified using cut-off values of 20 mm of mercury (both negative and positive). The distribution of blood pressure changes was significantly correlated with the distribution of response, with correlations ranging from 0.46 to 0.62 for males and 0.40 to 0.62 for females (Ward et al., Altman et al.,). For those males whose blood pressure response was less than −20 mm, the actual change in blood pressure was a drop of approximately 10 mm of mercury for both diastolic blood pressure and systolic blood pressure. Conversely, for males where the response was greater than 20 mm, the blood pressure increased by approximately 20 mm of mercury in migrants, but only 10 mm of mercury in nonmigrants. The pattern was somewhat more complex for females. Nonmigrant females with a response less than −20 had drops in blood pressure of between 5 and 6 mm for systolic blood pressure and 8 to 11 mm for diastolic blood pressure. However, migrant females with a response greater than 20 mm had a blood pressure rise of nearly 35 mm systolic pressures, compared with 16 mm for nonmigrants. Diastolic pressures in migrant females showed a rise of about 9 mm, compared with 18 mm for nonmigrants. Hence, it appears that there is considerable variability in the blood pressure response, and that this is more variable than blood pressure change alone.

We also defined the standardized response (blood pressure response standardised by the changed relationships between concomitant variables and blood pressure) to identify the magnitude of the response due to unknown factors in the new environment. The distribution of these variables, which is reported elsewhere, suggests that a variety of sociocultural factors are involved in influencing the magnitude of blood pressure response (Altman et al.,). In Table 3 we give the comparison between blood pressure change, the full response, and the standardized response for migrants and nonmigrants, respectively. The values for the full response are similar to the average values in Table 2 but the standardized response tends to be close to zero. The actual relationship between the full response and the standardized response changes according to sex and migrant status. As expected, after controlling for the changed relationships between concomitants and blood pressure (relative to the relationship defined in the premigrant population), the nonmigrants have a standardized response score close to zero, suggesting that "new" factors have not emerged in the Tokelau environment. By contrast, the standardized response score in migrants tends to be of the order of −2 mm for males and between 0.5 mm and 1.5 mm in females, suggesting that new factors play some slight role in New Zealand. However, the larger difference between the full response and the

TABLE 3. Change in blood pressure in Tokelau—comparison of mean blood pressure change, full response and standardised response, for migrants and nonmigrants

		Migrants			Nonmigrants		
		Blood pressure change	Response: Full	Response: Standardised	Blood pressure change	Response: Full	Response: Standardised
Males							
	Systolic	6.45	0.19	−2.46	−1.47	−5.64	0.28
	Diastolic	5.97	3.85	−1.97	−0.69	−3.67	−0.21
Females							
	Systolic	3.46	−2.77	−0.48	−1.20	−6.52	−0.28
	Diastolic	−0.59	−2.57	−1.62	−2.27	−5.48	−0.44

standardized response indicates that the impact of the changed relationships between concomitants and blood pressure is a more important determinant of blood pressure change in migrants than the influence of new factors.

ANALYSING COMPONENTS OF VARIATION IN BLOOD PRESSURE RESPONSE

Having defined the variable of interest, we now wish to examine whether familial aggregation is an important factor in influencing the distribution of response in either the migrant or nonmigrant populations. If familial aggregation is found to be present, it will be pertinent to see if a pattern of transmission within families is consistent with a genetic model. While there are a variety of statistical techniques that could be employed for this, including standard segregation analysis, path analysis, and the calculation of "indices of segregation" (Karlin et al., 1979), we have elected to use the mixed model of Morton and McLean (1974) for a variety of reasons. First, the application of this model permits the identification of the existence of more than one component in the distribution of response scores. Second, as recently pointed out by Lalouel et al., incorporation of transmission probabilities into the mixed model allows explicit tests of a genetic hypothesis. While the application of the mixed model is worthy of extended discussion, space does not permit this here. A good overview is given by Elston (1979, 1981). The specific program that we have used is a modification of the "pointer" program of Lalouel and Morton (1981). In applying this model to the data set, we have first transformed the standardized data into a normal variate with a mean of zero and a variance of 1. In this data, excess skewness did not appear and a power transformation was unnecessary. In applying the mixed model, we have set up a number of specific tests, as recommended by Elston (1981).

The first test is to evaluate whether there is more than one distribution in the population of interest. This involves defining the deviation of the mean of the minor component from the mean of the major component in terms of standard deviation units, as well as the frequency of the minor component. Second, conditional on the existence of the appropriate number of distributions, we test for the existence of polygenic inheritance. Last, conditional on these results, explicit genetic hypotheses are tested. This requires estimation of the degree of dominance, the gene frequency of the presumptive genetic trait, and the three transmission parameters. For a genetic model to be accepted, the transmission parameters must be consistent with Mendelian expectation. The overall results of carrying out these steps are presented in Table 4, for both migrants and nonmigrants.

For all measures of response, the null hypothesis of only a single distribution was rejected in favour of the alternate hypothesis that two distributions were present. In both migrant and nonmigrant populations, the existence of a minor component contributed significantly to the overall variability in blood pressure response. The proportion of the total variance attributed to the mixture of the minor component with the major component was approximately 30% for systolic response in both populations and for diastolic response in migrants, but was only 17% for diastolic response in nonmigrants (Table 4). It will be noted that the mean of the minor component exhibits a substantial deviation from the mean value of the major component—approximately three to four standard deviation units. Hence, this analysis suggests there is a small number of individuals in each population who have a blood pressure response substantially in excess of the overall response displayed by the rest of the population. This raises the question of whether these high responders aggregate in families, and if so, whether a single Mendelian locus is likely to be responsible.

Before attempting to identify the possible effect of a major locus, it is desirable to test for the existence of polygenic inheritance, conditional on the existence of two distributions. In this instance, the likelihood that a polygenic component contributes to the variability about the mean of each of the two constituent distributions was significantly greater than the likelihood that the dispersion about the two means was entirely due to environmental factors. The maximum likelihood estimates of the contribution of the polygenic component to the total variability are given in the final column of Table 4. These values, which vary from 18% to 28%, are similar in overall magnitude to the proportion of the total variance attributable to the existence of admixed distributions. Together, the combined influence of the presence of two components, plus the existence of a well-defined polygenic component, accounts for half the total variability observed for blood pressure response in these populations.

Finally, the existence of familial clustering of high responders was examined, along with estimation of the appropriate genetic parameters (degree of dominance and frequency of the putative allele causing the high response). As Lalouel et al. emphasise, the underlying cause of familial clustering of the minor component is most stringently defined by critically evaluating the transmission parameters. For Mendelian inheritance, these three probabilities must take on the values 1.0, 0.5, and 0.0, respectively. If the likelihood associated with these parameter values is significantly less than the likelihood associated with an alternate set of transmission probabilities (not compatible with Mendelian segregation), then the hypothesis that the mixture of several components in the distribution is due to a major gene should be rejected.

When this strategy was applied to the analysis of blood pressure response, three distinct results obtained. First, in the nonmigrant population there was no evidence for familial clustering for either systolic or diastolic response. For each type of response in this population, the probability of transmitting the major ("normal") component to an offspring was simply the population frequency, irrespective of the putative genotype of the parent. Thus, in the nonmigrant population, individuals with high values of blood pressure response do not cluster in nuclear families but are randomly distributed throughout the whole population.

By contrast, in the migrant population two types of familial clustering of blood pressure response were observed. For diastolic response, there is significant aggregation within nuclear families, but the estimated values for the transmission parameters are not consistent with those expected under a Mendelian model. In particular, parents who are high responders tend to have significantly more offspring with normal response values than would be predicted by a Mendelian model. Nevertheless, since there is a marked tendency for high responders to cluster into families, the underlying cause may be cultural rather than genetic.

Last, for systolic response in migrants, high responders not only cluster within families, but the likelihood associated with the parameters

TABLE 4. Characteristics of blood pressure response in Tokelau: analysis by the "mixed model"

	Number of distributions	Analysis of mixtures						Variance components: % total variance due to	
		Transmission of minor component							
		τ_1	τ_2	τ_3	Interpretation	Frequency	Deviation from major component	Mixture of distributions	Polygenic inheritance
Migrants									
Systolic	2	1.00	0.50	0.00	Familial ? Mendelian (recessive)	0.201	2.74	29.2	17.6
Diastolic	2	0.95	0.68	0.25	Familial non-Mendelian	0.072	4.14	29.5	22.3
Nonmigrants									
Systolic	2	0.80	0.80	0.80	Nonfamilial	0.198	2.83	30.1	20.8
Diastolic	2	0.88	0.88	0.88	Nonfamilial	0.122	3.42	17.1	27.5

that determine a recessive model is not significantly different from the likelihood associated with the general solution. Hence, the results are consistent with the hypothesis that in the migrant population, the minor component of high responders is due to the effect of a major gene segregating as a recessively inherited phenotype. While this is provocative, more work is needed to confirm this finding. As indicated in Table 4, at this stage we only tentatively attribute the distribution of high systolic response values in the migrant population to the action of a major gene.

This brief overview of the analysis of components of variability in blood pressure response indicates that polygenic factors and the mixture of two components contribute about equally to the total variance of blood pressure response. Also, in addition to the familial aggregation of blood pressure response attributable to a polygenic component in the migrant population, familial aggregation is also influenced by marked familial clustering of the minor component. In the case of systolic response, this distribution of the minor component within families is consistent with the hypothesis that the underlying cause is a major gene displaying recessive inheritance. For diastolic response, the estimated transmission probabilities are not consistent with the genetic model. This suggests that alternative, nongenetic factors, such as sociocultural influences, may be responsible for the observed levels of familial aggregation in diastolic response.

DISCUSSION

The above analysis of blood pressure response gives some insight into the impact of major environmental change on the distribution of an important cardiovascular risk factor. While simple comparisons of blood pressure levels between migrants and nonmigrants had revealed extensive differences between these two populations (Beaglehole et al., 1977a,b; Ward et al., 1980b; Ward and Prior, 1980), this was difficult to interpret in terms of response to environmental change. In particular, comparison of both migrant and nonmigrant blood pressures with the "premigrant" levels indicated that nonmigrant blood pressures had dropped in the Tokelau environment over the 5–8-year period separating the two surveys. By contrast, migrant levels had risen slightly. This was puzzling and difficult to interpret. It was not clear if these apparent changes in blood pressures were due to specific shifts in the distribution of concomitant factors such as obesity, or if other factors were influencing blood pressure in the new environment. The difficulty of interpreting these results was compounded by the lack of a complete longitudinal survey for all members of the population.

As described above, the construction of a generalized response variable eliminates many of the procedural problems in evalvating blood pressure change over time, when complete longitudinal information is not available. More importantly, it allows assessment of the relative role played by specified concomitants in influencing blood pressure change. If the observed differences in blood pressure between migrants and nonmigrants were entirely due to specific changes in weight, percent fat, age, and other concomitants (with the relationship between blood pressure and concomitants being the same in both populations) this would be of little genetic interest. If the changes in blood pressure are attributable solely to changes in concomitant factors, the proximate cause of blood pressure change may then be classified as environmental in origin. The only exception to this is if the concomitants themselves are genetically determined, in which case a more complex model of analysis, involving a definition of correlated response, is required.

Analysis of blood pressure response in the Tokelau situation indicates that the differences in blood pressure between migrants and nonmigrants are attributable to changes in the relationship between blood pressure and concomitants, plus the addition of new concomitants, as well as changes in the concomitants. The importance of this finding can be seen by examination of Table 3. Relative to the premigrant population, the age-standardized difference between migrants and nonmigrants is given by the difference of blood pressure change. For males this is 7.9 mm for systolic blood pressure and 6.7 mm for diastolic pressure, while for females this is 4.7 mm for systolic pressure and 1.7 mm for diastolic blood pressure.

Hence, for both males and females the difference between the full response for systolic values is less than the total difference in blood pressure. This indicates that changes in concomitants have made some contributions to the observed differences in blood pressure between the two populations. For males, the change in concomitants contributes 26.4% of the observed difference, whereas for females it contributes 17.4%. However, the major portion of the observed difference in blood pressures between migrants and premigrants is due to the changed relationship between blood pressure and concomitants which accounts for 73.6% of the difference in males and 82.6% in females.

For diastolic pressures, the magnitude of the difference between full response is greater than the observed difference between blood pressure levels in the two populations. This indicates that, although the distribution of the relevant concomitants has changed in the two populations, the direction of change has been reversed. The nonmigrants have higher expected values of blood pressure on the basis of their observed concomitant factors compared with the expected values for the migrants. Hence, for diastolic pressures, the observed differences in blood pressure are entirely due to the changed relationships between concomitants and blood pressure. The impact of this changed relationship would be even more marked had not the distribution of concomitants also shifted in the two populations.

The influence of new concomitants on the distribution of blood pressure is measured by comparing the difference in standardized response. As Table 3 indicates, the nonmigrants have higher values than migrants. This indicates that new concomitant factors may have entered the relationship since migration. However, these new factors, which are presumed to be sociocultural in origin (such as smoking habits, alcohol use, etc.—Altman et al.,), have tended to raise blood pressure more in Tokelau than in New Zealand. Although the overall effect is small, this is a surprising finding and needs further investigation.

Finally, the role played by genetic-environmental interactions in the raised blood pressure following migration can be deduced from the genetic analysis of the response variable. The response variable measures the consequences of migration and attendant sociocultural change in such a way that it incorporates any component due to genetic-environmental interaction. (A response score of zero suggests that genetic-environmental interaction plays no role in the blood pressure change, though the converse does not apply.) An apparent genetic basis for the observed distribution of response scores would indicate the existence of a genetic-environmental interaction influencing the observed changes in blood pressure.

As summarized in Table 4, there is evidence for high responders in both the migrant and nonmigrant population. The present analysis suggests that these individuals, relatively few in number, are randomly distributed throughout the nonmigrant population but tend to cluster within families in the migrant population. The existence of familial clustering in migrants raises the possibility that the high responders may be genetically distinct from the rest of the population. In the case of systolic blood pressure response, the results are consistent with the hypothesis that a major locus, displaying recessive inheritance, may be involved. However, since this does not appear to be the case for the class of high responders for diastolic blood pressure, some caution should be used in interpreting these results. In particular, it is possible that sociocultural factors also play a role in the observed distribution of high-responding individuals, and this may have given rise to the observed degree of familial clustering.

Our previous analysis of transmission parameters (Ward and Prior, 1980) indicated that the patterns of familial transmission of sociocultural factors were quite different in the migrant and nonmigrant populations. Thus, our subsequent analyses will need to take account of the familial distribution of sociocultural factors as well as the possible impact of a major gene. The above results indicate that both factors may underlie the observed changes in the distribution of blood pressure and other cardiovascular risk factors in the migrants, with sociocultural factors playing the dominant role. This agrees with our previous observations on the pattern of transmission of familial background. Hence, even though it is likely that genetic-environmental interactions play a role, careful analysis of the familial distribution of sociocultural factors will be needed. The present analysis lends weight to the conclusion that sociocultural factors have played a far more important role than genetic factors in determining the increased frequency of hypertension observed after migration. However, genetic susceptibility cannot be entirely discounted since there is still some suggestive indications that both polygenic and major loci may play some role in the blood pressure changes.

ACKNOWLEDGMENTS

This research was supported by award HL 26869 from the Heart, Lung and Blood Institute, National Institutes of Health, USA.

LITERATURE CITED

Abrahams, DG, Alele, CA, and Barnard, B (1960) The systematic blood pressure in a rural West African Community. West African Med. J. *9*:45.

Akinkugbe, OO (1969) Hypertensive disease in Ibadan, Nigeria. East African Med. J. *46*:313.

Akinkugbe, OO, and Ojo, OA (1969) Arterial pressures in rural populations in Nigeria. Brit. Med. J. *2*:222–224.

Akinkugbe, OO (1976) Epidemiology of Hypertension and Stroke in Africa. In S Hateno et al (eds): Hypertension and Stroke Control in the Community. Geneva: World Health Organisation, pp. 28–42.

Altman, N, Ward, RH, Prior, IAM Patterns of blood pressure response in migrant and nonmigrant Tokelauans. II. Factors influencing blood pressure response in adults. J. Chron. Dis. (submitted).

Beaglehole, R, Salmond, CE, Prior, IAM (1975) A family study of blood pressure in Polynesians. Int. J. Epidem. *4*:217–220.

Beaglehole, R, Salmond, CE, Eyles, EF (1977a) A longitudinal study of blood pressure in Polynesian children. Am. J. Epidem. *105*:87–89.

Beaglehole, R, Salmond, CE, Hooper, A, Huntsman, J, Stanhope, JM, Cassel, JC, Prior, IAM (1977b) Blood pressure and social interaction in Tokelau migrants in New Zealand. J. Chron. Dis. *30*:803–812.

Beaglehole, R, Eyles, E, Salmond, C, Prior, IAM (1978) Blood pressure in Tokelauan children in two contrasting environments. Am. J. Epid. *108*:283–288.

Beaglehole, R, Eyles, E, Prior, IAM (1979) Blood pressure and migration in children. Int. J. Epid. *8*:5–10.

Bechgaard, P (1946) Arterial hypertension. A follow-up study of 1,000 hypertonics. Acta. Med. Scand. [Suppl.] *172*:1–158.

Canessa, M, Adragna, N, Solomon, H, Connolly, T, Toteson, D (1980) Increased sodium-lithium countertransport in red cells of patients with essential hypertension. N. Engl. J. Med. *302*:772–775.

Cassel, J (1974) Hypertension and cardiovascular disease in migrants: A potential source of clues. Int. J. Epidemiol. *104*:107–123.

Davis, D, and Klainer, MJ (1940) Studies in hypertensive heart disease. I. The incidence of coronary atherosclerosis in cases of essential hypertension. Am. Heart. J. *19*:185–203.

Donnison, CP (1929) Blood pressure in the African native: Its bearing upon aetiology of hyperplesia and arteriosclerosis. Lancet *1*:6.

Dreyfuss, W, Hamosh, P, Adam, YG, and Kallner, B (1961) Coronary heart disease and hypertension among Jews emigrated to Israel from the Atlas mountain region of North Africa. Am. Heart. J. *62*:470–477.

Elston, RC (1979) Major locus analysis for quantitative traits. Am. J. Hum. Genet. *31*:655–661.

Elston, RC (1981) Segregation analysis. In: Advances in Human Genetics. *13*:62–121.

Epstein, FH, Eckhoff, RD (1967) The epidemiology of high blood pressure—Geographic distribution and etiological factors. In J Stamler, R Stamler, and TM Pullman (eds): The Epidemiology of Hypertension. New York: Grune and Stratton, pp. 155–166.

Feinlieb, M, Garrison, RJ, Havlik, RJ (1980) Environmental and genetic factors affecting the distribution of blood pressure in children. In RM Lauer and RB Shekelle (eds): Childhood Patterns of Arthrosclerosis and Hypertension. New York: Raven Press, pp. 271–279.

Gampel, B, Slome, C, Scotch, N, Abramson, JH (1962) Urbanisation and Hypertension among Zulu Adults. J. Chron. Dis. *15*:67–70.

Garay, RP, Dagher, G, Pernollet, MG, Devynck, MA, Meyer, P (1980) Inherited defect in a Na^+, Ka^+ co-transport system in erythrocytes from essential hypertensive patients. Nature. *284*:281–283.

Haldane, JBS (1946) The interaction of nature and nurture. Ann. Eugen. *13*:197–205.

Johnson, TO (1971) Arterial blood pressures and hypertension in an urban African population sample. Br. J. Prev. Soc. Med. *25*:26–33.

Kannel, WB, Schwartz, MJ, and McNamara, PM (1969) Blood pressure and risk of coronary heart disease: The Framingham study. Dis. Chest *56*:43–58.

Karlin, S, Carmelli, D, and Williams, R (1979) Index measures for assessing the mode of inheritance of continuously distributed traits. I. Theory and justifications. Theor. Pop. Biol. *16*:81–106.

Keith, KM, Wagener, HP, Barker, NW (1939) Some different types of essential hypertension—Their course and prognosis. Am. J. Med. Sci. *197*:332–339.

Kempthorne, O (1978) Logical, epistemiological and statistical aspects of nature-nurture data interpretation. Biometrics. *34*:1–23.

Keys, A (ed) (1970) Coronary Heart Disease in Seven Countries. Circulation *41[Suppl. 1]*:1–211.

Lalouel, JM, Morton, NE (1981) Complex segregation analysis with pointers. Hum. Hered. *31*:312–321.

Lalouel, JM, Elston, RC, Morton, NE manuscript in preparation.

Levitt, EN et al (1974) A study of hypertension in the Zambian African. East Afr. Med. J. *51*:869–877.

Maddocks, I (1961) Possible absence of essential hypertension in two complete Pacific island populations. Lancet *2*:396–398.

Marmot, MG (1979) Epidemiological Basis for the Prevention of Coronary Heart Disease. Bull. WHO *57*:331–347.

Morton, NE, and MacLean, CJ (1974) Analysis of family resemblance. III. Complex segregation of quantitative traits. Am. J. Jum. Genet. *26*:489–503.

Murphy, W (1955) Some observations on blood pressure in the humid tropics. N. Z. Med. J. *54*:64–67.

Neel, JV (1962) Diabetes mellitus: A "thrifty" genotype rendered detrimental by "progress"? Am. J. Hum. Genet. *14*:353–362.

Neilson, G, Williams, G (1978) Blood pressure and valvular and congenital heart disease in Torres Strait islanders. Med. J. Aust. [Spec. Suppl.] *1*:12–16.

Oliver, WJ, Cohen, EL, Neel, JV (1975) Blood pressure, sodium intake and sodium related hormones in the Yanomama Indians, a "no-salt" culture. Circulation *52*:146–151.

Oliver, WJ (1980) Sodium homeostasis and low blood pressure populations. In H Kesteloot and JV Joossens (eds): Epidemiology of Arterial Blood Pressure. The Hague: Nijhoff.

Page, LB, Damon, A, Moellering, RC (1974) Antecedents of cardiovascular disease in six Solomon Island societies. Circulation *49*:1132–1146.

Page, LB (1980) Dietary sodium and blood pressure: Evidence from human studies. In R Shekelle and RM Lauer (eds): Childhood Prevention of Atherosclerosis and Hypertension. New York: Raven Press.

Pickering, G (1967) The inheritance of arterial pressure. In J Stamler, R Stamler and TN Pullman (eds) The Epidemiology of Hypertension. New York: Grune and Stratton, pp. 18–27.

Prior, IAM (1970) WHO Report on Cardiovascular Epidemiology in the Pacific. Geneva: WHO, pp. 28.

Prior, IAM, Harvey, HPB, Neave, MN, Davidson, F (1966) The Health of Two Groups of Cook Island Maoris. Wellington: New Zealand Department of Health.

Prior, IAM, Stanhope, JM, Evans, JG, Salmond, CE (1974) The Tokelau Island Migrant Study. Int. J. Epidemiol. *3*:225–232.

Prior, IAM, Hooper, AB, Huntsman, JW, Stanhope, JM, Salmond, CE (1977) The Tokelau Island Migrant Study. In GA Harrison (ed): Population Structure and Human Variation. Cambridge: Cambridge University Press, pp. 165–185.

Prior, IAM, Evans, JG, Harvey, HPB, Davidson, F, Lindsey, M (1968) Sodium intake and blood pressure in two Polynesian populations. N. Engl. J. Med. *279*:515–520.

Prior, IAM, Stanhope, JM (1980) Blood pressure patterns, salt use and migration in the Pacific. In H Kesteloot and JV Joossens (eds): Epidemiology of Arterial Blood Pressure. The Hague: Nijhoff. pp. 243–262.

Schrire, V (1958) The renal incidence of heart disease at Groote Schuur, Hospital, Capetown: hypertension and valvular disease of the heart. Amer. Heart J. *56:*742–751.

Scotch, NA (1963) Sociocultural Factors in the Epidemiology of Zulu Hypertension. Am. J. Public Health *53:*1205–1213.

Shaper, AG (1967) Blood pressure studies in East Africa. In J Stamler, R Stamler and TN Pullman (eds): The Epidemiology of Hypertension. New York: Greene and Stratton, pp. 139–149.

Sinnett, PF, and Whyte, HM (1973) Epidemiological studies in a total highland population, Tukisenta, New Guinea. Cardiovascular disease and relevant clinical, electrocardiographic, radiological and biochemical findings. J. Chron. Dis. *26:*265.

Stamler, J (1980) Improved Life Styles: Their potential for the primary prevention of atherosclerosis and hypertension in childhood. In RM Lauer and RB Shekelle (eds): Childhood Prevention of Atherosclerosis and Hypertension. New York: Raven Press, pp. 3–36.

Stamler, J, Rhomberg, P, Schoenberg, JA et al (1975) Multivariate analysis of the relationship of seven variables to blood pressure. J. Chron. Dis. *28:*527–548.

Stanhope, JM, Prior, IAM (1976) The Tokelau Island migrant study: Prevalence of various conditions before migration. Int. J. Epidemiol. *5:*259–266.

Stanhope, JM, Sampson, VM, and Prior, IAM (1981) The Tokelau Island Migrant study: Serum lipid concentrations in two environments. J. Chron. Dis. *34:*45–55.

Turner, HN, and Young, SSY (1969) Quantitative gentics in sheep breeding. Cornell Univ. Press. New York.

Vaughan, JP, Miall, WE(1979) Cardiovascular Measurements in Subjects of African Origin. Bull. WHO *57:*281–289.

Vint, FW (1937) Post mortem findings in natives in Kenya. East Afr. Med J. *13:*332.

Ward, RH, Chin, PG, Prior, IAM (1979) Genetic epidemiology of blood pressure in a migrating isolate: Prospectus. In CF Sing, and MH Skolnick (eds): Genetic Analysis of Common Diseases. New York: Alan R. Liss, pp. 675–709.

Ward, RH, Raspe, PD, Ramirez, ME, Kirk, RL, Prior, IAM (1980a) Genetic structure and epidemiology: The Tokelau Island study. In Eriksson (ed): Population Structure and Genetic Disorders. London: Academic Press, pp. 301–325.

Ward, RH, Chin, PG, Prior, IAM (1980b) The Effect of Migration of the Familial Aggregation of Blood Pressure. Hypertension *2* [Suppl. 1]:43.

Ward, RH, Prior, IAM (1980) Genetic and sociocultural factors in the response of blood pressure to migration of the Tokelau population. Med. Anthropol. *4:*339–366.

Ward, RH, Altman, N, Prior, IAM Patterns of blood pressure response in migrant and nonmigrant Tokelauans. I. Defining response. J. Chron. Dis. (submitted).

Whyte, HM (1958) Body fat and blood pressure of natives in New Guinea; reflections on essential hypertension. Aust. Ann Med. *7:*36–42.

AMERICAN JOURNAL OF PHYSICAL ANTHROPOLOGY 62:107–114 (1983)

Diabetes Mellitus in the Pima Indians: Genetic and Evolutionary Considerations

WILLIAM C. KNOWLER, DAVID J. PETTITT, PETER H. BENNETT, AND ROBERT C. WILLIAMS
National Institute of Arthritis, Diabetes, and Digestive and Kidney Diseases, Phoenix, Arizona 85014 (W.C.K., D.J.P., P.H.B.) and Arizona State University Tempe, Arizona 85281 (R.C.W.)

KEY WORDS Diabetes mellitus, Obesity, American Indians, Genetics

ABSTRACT Non-insulin-dependent diabetes mellitus is a common disease in the Pima Indians. It is familial and strongly related to obesity. Neel (1962) suggested that the introduction of a steady food supply to people who have evolved a "thrifty genotype" leads to obesity, insulin resistance, and diabetes. Our findings in the Pimas of differences in insulin sensitivity in different metabolic pathways suggest that the thrifty genotype involves the ability of insulin to maintain fat stores despite resistance to glucose disposal. The recent increase in diabetes incidence following the availability of an abundant food supply suggests that the ability to store energy efficiently during cycles of feast and famine may now lead to obesity, insulin resistance, and diabetes.

The ancestors of the Pima Indians of the Gila River Indian Community of central Arizona are thought to have lived near the Gila River for about 2,000 years. According to Haury (1976) the Pimas descended from the Hohokam, who moved into the Gila River Valley from Mexico around 300 B.C.

The Hohokam lived successfully in their desert environment by hunting and gathering food and by building an elaborate irrigation system diverting water from the Gila River through canals, many of which are still in use today. The Pimas' successful adaptation to desert life continued until this century, when increasing settlement of the area by European-derived people led to diversion of the Pimas' water supply and disruption of their traditional agricultural lifestyle.

Now the Pima community shares many economic and social attributes with non-Indian American society. Although irrigation agriculture remains a major industry, most food is purchased. The Pimas are now plagued by several chronic diseases which may be related to their history. Obesity and diabetes are two such problems which are the subject of this paper. The investigation of the development and natural history of obesity and diabetes has been approached through a longitudinal epidemiologic study (Bennett et al., 1976). Since 1965, Pimas at least 5 years of age and living in the Gila River Indian Community of Central Arizona have been examined periodically for the presence of diabetes and obesity. Many subjects have been examined before and after the onset of diabetes, facilitating study of the prediabetic state. In addition, Pima and Caucasian subjects have been admitted to a metabolic research unit for more detailed studies of insulin and carbohydrate metabolism.

Obesity, defined as an excess of adipose tissue, cannot readily be measured directly in large-scale studies, so it is usually approximated by some index based on height and weight. We use the body mass index, computed as body weight divided by the square of the height, expressed in units of kg/m^2 (Khosla and Lowe, 1967). Figure 1 shows the mean body mass index as a function of age in Pima men and women compared with men and women in the general U.S. population as estimated by the Health and Nutrition Examination Survey. At all ages the Pimas are more obese, but especially in the younger adult years. These

Received August 5, 1982; accepted March 16, 1983.

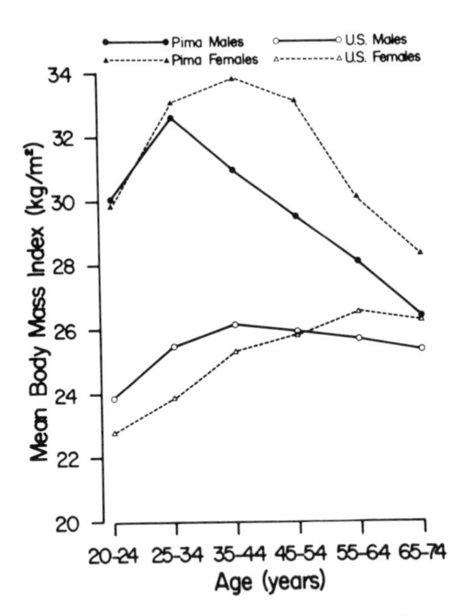

Fig. 1. Mean body mass index (weight/height^2 in kg/m^2) in Pima Indian adults, compared with measurements in the general U.S. population from the Health and Nutrition Examination Survey, 1971–1974. Reprinted with permission from Knowler et al. (1981).

high mean body mass indices are not just due to a few extremely obese individuals. At ages 20–34 years, over 90% of Pima men and women exceed the U.S. age-specific medians of body mass index, and over half of the Pimas exceed the U.S. 90th percentiles (Knowler et al., 1981).

Although measurement data were sparse before our studies began in 1965, obesity was reportedly rare as recently as a few decades ago. We compared mean weights for height in Pima children aged 5–18 years examined in the present studies with measurements of 205 Pima children judged to be aged 18 years or younger by Hrdlička (1908). In Figure 2, mean weights are given for categories of height rather than age because precise ages were not reported for Hrdlička's subjects. Pima children now are heavier, by up to about 10 kg on average for the tallest boys, than Pima children at the same height at the beginning of this century. This suggests that the degree of obesity in the Pimas has increased since that time.

The Pimas have an extremely high prevalence of diabetes, shown in Figure 3 as a function of age. Approximately 50% of adults over the age of 35 years have diabetes. As in other populations, the disease has a much lower frequency in children.

CLASSIFICATION OF DIABETES MELLITUS

Diabetes is diagnosed by an excessively high concentration of glucose in the blood occurring spontaneously or following an oral glucose challenge (National Diabetes Data Group, 1979). Most people with diabetes can be classified into one of two major types which are described in Table 1. Insulin-dependent (type I) diabetes is characterized by dependence on exogenous insulin to prevent ketoacidosis and death, by the presence of antibodies to pancreatic islet cells, and often by an abrupt onset of symptoms. Non-insulin-dependent (type II) diabetes is characterized by ketosis resistance, lack of islet-cell antibodies, and frequently an insidious or asymptomatic onset. There is a strong association with obesity. Although exogenous insulin may be required to control some disease manifestations, endogenous insulin secretion is preserved and insulin treatment is not essential for survival. In causal terms, the major distinction between the two types appears to involve insulin. In type I diabetes, the basic physiologic problem is lack of insulin. In type II diabetes, the basic physiologic problem appears to be resistance to the action of insulin, inappropriate insulin secretion, or a combination of the two. Because of the different basic defects in the two types and because of evidence from twin concordance and family studies suggesting different degrees of heritability and modes of inheritance (Barnett et al., 1981; Irvine et al., 1977), the two types of diabetes are generally considered to be causally distant. There may also be genetically distinct subtypes within these two major types.

Diabetes in the Pimas is almost exclusively non-insulin-dependent, or type II (Knowler et al., 1979), which is also the most common type of diabetes throughout the world. The remainder of this paper will be limited to type II diabetes.

INCREASING PREVALENCE OF DIABETES

The high prevalence rates of both obesity and diabetes are probably recent phenomena in Pima Indians. Temporal changes in diabetes prevalence are difficult to estimate because of changing diagnostic methods. However, in the medical literature of the early part of this century, diabetes in the Pima Indians was not prominently mentioned. Hrdlička (1908), in his studies of the Pimas during the first decade of the present century, reported only one case of diabetes in the population. Russell (1908), writing in the same year, did not mention diabetes in a list of 28 diseases affecting the Pi-

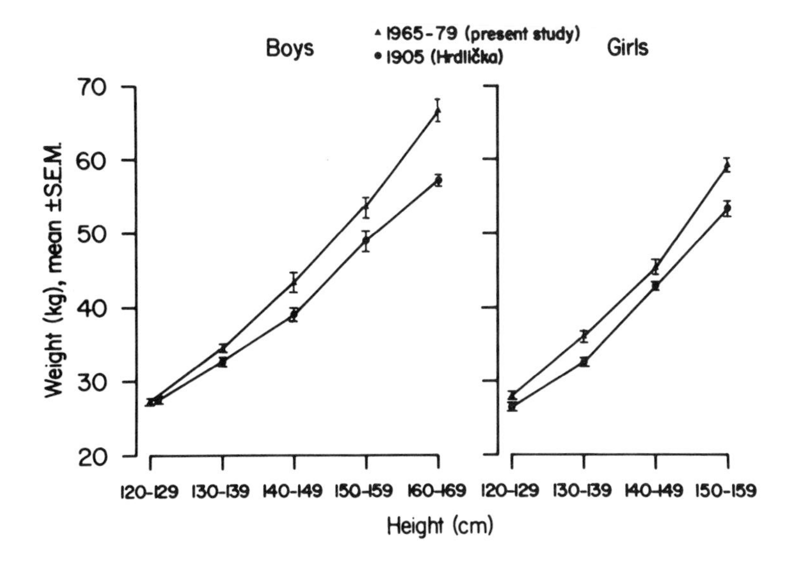

Fig. 2. Mean weights according to height in children measured by Hrdlička (1908) and in the present study.

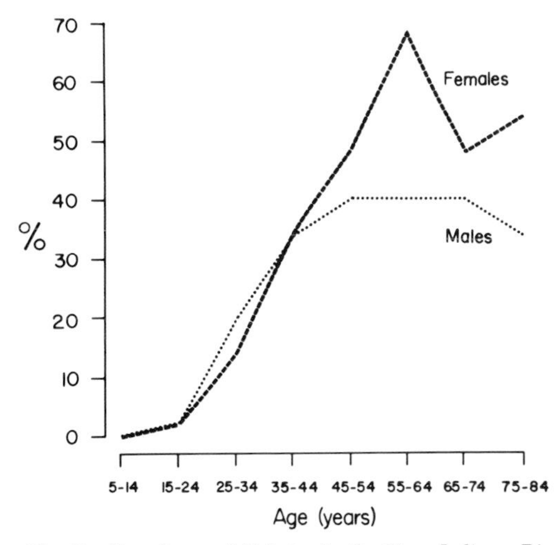

Fig. 3. Prevalence of diabetes in the Pima Indians. Diabetic subjects had a venous plasma glucose concentration of at least 200 mg/dl 2 hours after ingesting 75 gm of carbohydrate or had previous documentation of diabetes in their medical records. Reprinted with permission from Knowler et al. (1978).

mas. Even 32 years later, Joslin (1940), through a review of medical records, concluded that diabetes prevalence was similar in American Indians and in the general population. Only since the 1950s has diabetes been reported as having an excessive frequency in the Pimas (Cohen, 1954; Parks and Waskow, 1961). Now it is a major health problem, affecting half of the adults at least 35 years old. The disease is not only frequent but serious, often accompanied by vascular disease typical of diabetes in other populations. There is also evidence for similar increases in diabetes prevalence over time in several other populations which, in the last few decades, have undergone major socioeconomic changes accompanied by increasing obesity. Such changes have occurred in many other American Indian tribes (West, 1978), Polynesians (Prior et al., 1978), and inhabitants of several Pacific Islands, such as Nauru (Zimmet, 1979). Migration may also be associated with increased diabetes prevalence (Cohen, 1961; Kawate et al., 1979). These observations suggest that there must be environmental, as well as hereditary, determinants of type II diabetes.

Although type II diabetes has long been regarded as a disease with a strong hereditary basis, understanding its genetics has been hampered by problems of diagnosis, classification, and environmental risk factors. Because the disease may be asymptomatic, it often remains undiagnosed. Furthermore, diabetes often first appears late in life, and many people with a diabetes-susceptibility genotype must escape the disease throughout their lives—i.e., penetrance is incomplete.

Diabetes incidence is the rate at which new cases of the disease develop in previously un-

TABLE 1. Distinction between the two major types of diabetes mellitus

	Insulin-dependent (Type I)	Non-insulin-dependent (Type II)
Dependent on exogenous insulin to prevent ketoacidosis and death	Yes	No
Antibodies to pancreatic islet cells and islet cell destruction	Yes	No
Onset of symptoms	Sudden	Insidious or asymptomatic
Endogenous insulin	Low or absent	Present
Basic physiologic problem	Insulin lack	Insulin resistance or inappropriate secretion

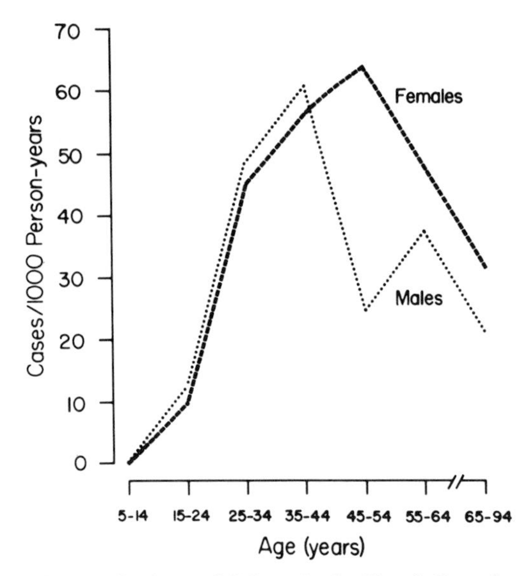

Fig. 4. Incidence of diabetes in the Pima Indians. Reprinted with permission from Knowler et al. (1978).

affected individuals. Age-sex-specific incidence rates are shown in Figure 4. The rates were highest between the ages of 25 and 55 years and were lower in older people. This finding is different from most other populations in which the incidence continues to rise with age. The age pattern in the Pimas facilitates study of the genetics of type II diabetes because the disease appears earlier than in other populations. The diabetes incidence rate, age-sex adjusted to the U.S. white population, is 27 cases/1,000 person-years, 19 times the diabetes incidence rate reported in the predominantly Caucasian population of Rochester, Minnesota (Knowler et al., 1978; Palumbo et al., 1976), and the highest reported diabetes incidence rate known to us.

FAMILIAL NATURE OF DIABETES IN THE PIMAS

As in other populations, diabetes in the Pimas is strongly familial. It is very common in the pedigree shown in Figure 5 and occurs in all four generations examined. This familial pattern is typical of many, but not all, Pima kindreds. The distribution of diabetes in this pedigree is consistent with an autosomal dominant mode of inheritance, but it is also compatible with other modes of inheritance, with or without additional environmental factors. Because of the strong evidence for environmental risk factors for diabetes in the Pimas (e.g., the increasing prevalence during this century, see above, and the effects of obesity and of diabetes in pregnancy, see below), a simple genetic hypothesis alone cannot explain the occurrence of the disease in the Pimas. Nevertheless, pedigree analysis, especially if genetic markers linked to diabetes can be identified, may be useful in understanding the mechanism.

The familial nature of diabetes is also demonstrated in Figure 6, which shows diabetes prevalence in offspring as a function of the diabetic status of the parents. Young men with two diabetic parents had considerably higher diabetes prevalence rates than those with only one diabetic parent or two nondiabetic parents. There were no cases of diabetes under age 30 in males with two nondiabetic parents. Similar results were found in Pima females.

A fundamental problem in diabetes genetics is that the prevalence rates, and hence inferences about inheritance, are so dependent on age. Many younger subjects could be misclassified in that they could have a "diabetic ge-

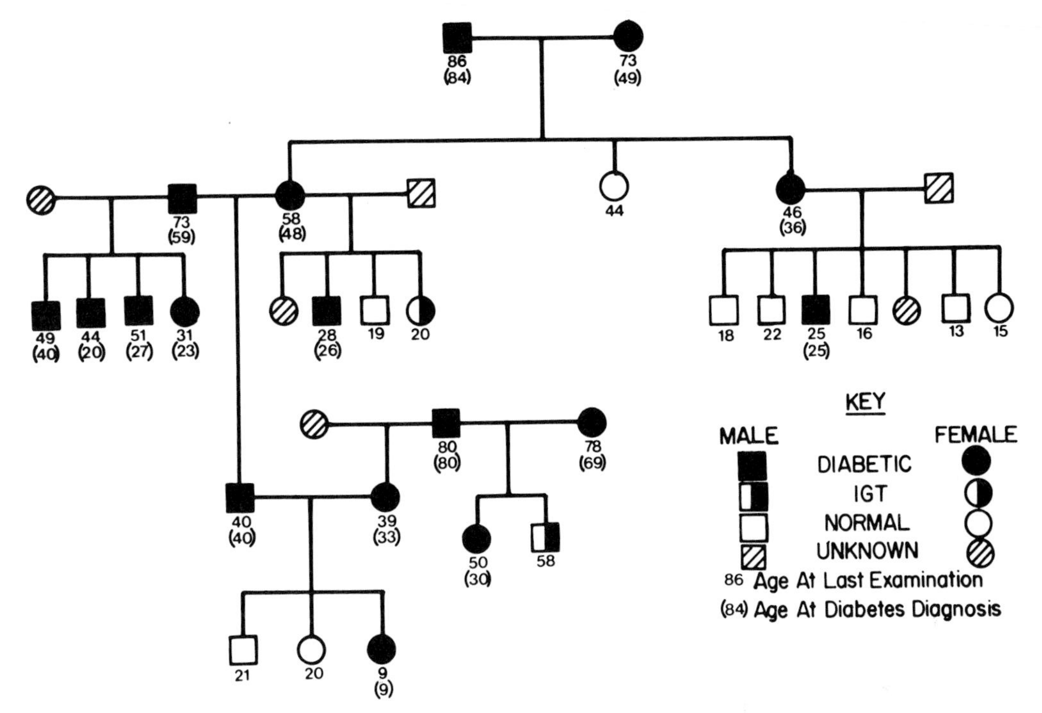

Fig. 5. Pedigree diagram showing diabetes in a Pima Indian kindred. The diabetic status of each subject was determined by a modified glucose tolerance test. Diabetics are defined as subjects with 2-hour postload plasma glucose concentrations at least 200 mg/dl. Subjects with impaired glucose tolerance had 2-hour glucose concentrations from 140 through 199 mg/dl, and those with normal glucose tolerance had 2-hour glucose concentrations less than 140 mg/dl. Reprinted with permission from Knowler et al. (1982).

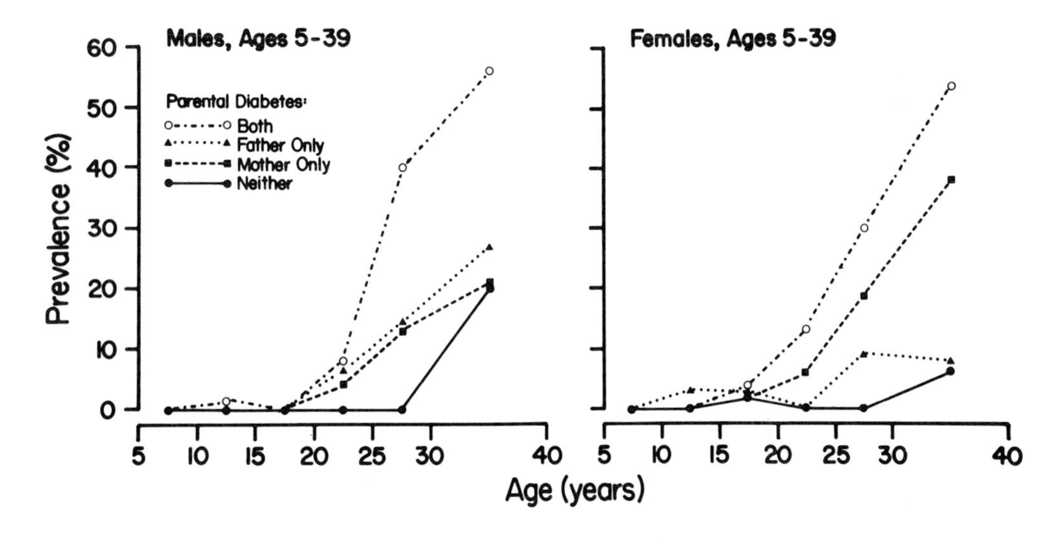

Fig. 6. Prevalence of diabetes in Pima Indians according to age, sex, and parental diabetes. Reprinted with permission from Knowler et al. (1982).

notype" but not yet express the disease, and the same problem probably occurs in some of the parents who may not express their "diabetic genotype."

Further evidence for a genetic component to diabetes in the Pimas comes from the association of an HLA antigen (HLA-A2) with the disease (Williams et al., 1981). Evidence for a gene-conferring susceptibility to type II diabetes has also been found in two other populations. Type II diabetes is associated with HLA-A2 in the Xhosa, a southern African black population (Briggs et al., 1980) and with HLA-Bw61 in Fiji Indians (Sarjeantson et al., 1981).

EFFECTS OF OBESITY AND PARENTAL DIABETES ON DIABETES INCIDENCE

Diabetes genetics is further complicated by obesity, a strong risk factor for type II diabetes. Obesity might be either an environmentally induced factor affecting the expression of a diabetes genotype or an early manifestation of a genetically determined obesity-diabetes syndrome.

Obesity has a marked effect on diabetes incidence. The age-sex adjusted incidence rates in Pimas are shown in Figure 7 as a function of the body mass index. The rates ranged from near zero in the thinnest individuals to 72 new cases/1,000 person-years in the most obese people. Although these results show a marked effect of obesity, the high incidence of diabetes in the Pimas cannot be attributed entirely to obesity. These incidence rates were compared with those of Rochester, Minnesota. For body mass indices between 20 and 25 kg/m^2, which are lower than means of the U.S. population in most age groups, the Pimas' age-sex adjusted incidence rate was 11 cases/1,000 person-years, or eight times as high as the 1.3 cases/1,000 person-years in Rochester (Knowler et al., 1981; Palumbo et al., 1976).

Obesity and parental diabetes are both strong diabetes risk factors which interact with each other. Age-adjusted diabetes incidence rates are shown in Figure 8 according to body mass indices and parental diabetes. Obesity and parental diabetes increase the diabetes incidence rate synergistically, with the highest rates in the obese subjects with two diabetic parents.

Diabetes in a woman during pregnancy puts her offspring at a higher risk of eventually developing diabetes. Among children aged 15–19 years, the diabetes prevalence was 0.6% in those whose mothers have never developed diabetes, and 1.4% in those born to nondiabetic mothers who subsequently developed diabetes, but 33% in those born after the onset of their mothers' diabetes (Pettitt et al., 1982). Thus in addition to having genetic determinants, the development of diabetes may also be influenced by the intrauterine environment.

Maternal diabetes during pregnancy is also a risk factor for obesity in the offspring, an effect which could not be explained by differences in the age or obesity of the mothers (Pettitt et al., 1981, 1983). Hence the propensity to obesity may be transmitted to offspring from

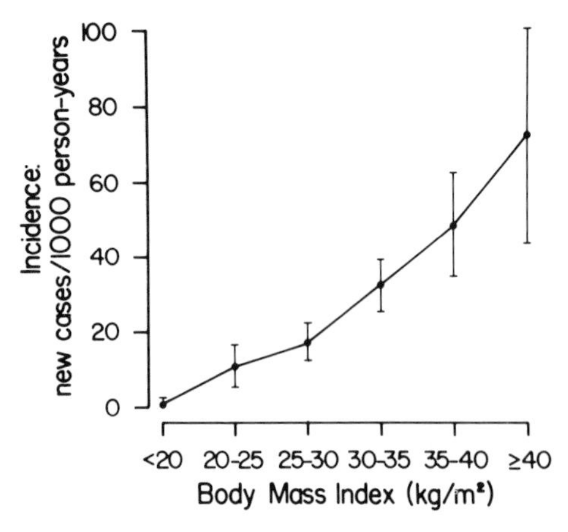

Fig. 7. Age-sex adjusted incidence rate (with 95% confidence intervals, i.e., ± 1.96 standard errors) for diabetes in Pima Indians, by body mass index. Reprinted with permission from Knowler et al. (1981).

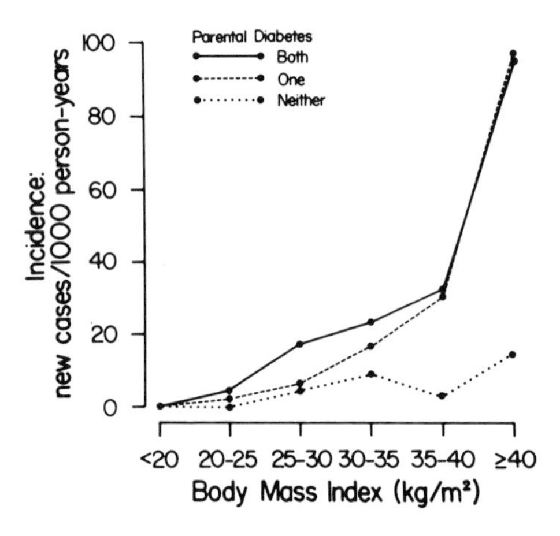

Fig. 8. Age-adjusted incidence of diabetes in Pima Indians according to body mass index. Rates are shown for subjects aged 5–44 years with two nondiabetic parents, parents discordant for diabetes, and two diabetic parents. Reprinted with permission from Knowler et al. (1981).

diabetic mothers, either genetically or via the intrauterine environment, and obesity in these offspring is in turn an important risk factor for diabetes. These findings illustrate the complexity of the relationship of obesity and diabetes and difficulties in making inferences about their inheritance.

HYPOTHESIS

It is interesting to speculate why the Pima Indians and several other populations which have recently undergone major socioeconomic changes have such a high incidence of obesity and of diabetes. Neel (1962) hypothesized that some isolated populations may have evolved what he called a "thrifty genotype" or ability to store calories in times of plenty. He argued that throughout most of human history, people have existed as hunters and gatherers in fluctuating states of feast or famine. The "thrifty genotype" should have had a selective advantage allowing an individual to store calories in the form of fat during times of feast and therefore be in a better position to survive the periods of relative famine. Neel suggested that this increased fat storage was accomplished by increased secretion of insulin in response to food. The introduction of a steady food supply to people who have evolved a "thrifty genotype" may lead to progressive obesity with high concentrations of insulin in the blood and impairment of insulin action, which may eventually result in diabetes. Hence in these circumstances the "thrifty genotype" may be detrimental.

Although Neel's hypothesis was proposed before our research on the Pimas was begun, our findings are consistent with the general principles which he outlined. If a "thrifty genotype" were to occur anywhere, it is not surprising that it would be found in these people who have subsisted for about 2,000 years by irrigation farming in the desert where the availability of water, and hence food, was intermittent. Neel recently suggested that the hypothesis needs to be revised in light of the distinction between type I and type II diabetes and the recent findings about insulin receptors and insulin sensitivity (Neel, 1982).

Metabolic studies comparing Pima Indians with Caucasians may provide clues as to what could mediate such a "thrifty genotype." Let us consider insulin secretion and insulin resistance, which may be the basic metabolic defects in type II diabetes. Insulin stimulates the utilization of blood glucose, thus lowering the blood glucose concentration, and it inhibits the breakdown of fat stores and the release of fat into the blood. Most people with diabetes or obesity are resistant to the action of insulin on glucose utilization (Nagulesparan et al., 1979; Reaven et al., 1976). Furthermore, in nondiabetic Pima Indians, insulin's ability to dispose of glucose is impaired in comparison with Caucasians (Nagulesparan et al., 1982). High concentrations of insulin in the blood are found in subjects with resistance to insulin action on glucose utilization. This hyperinsulinemia presumably compensates for the resistance, so that normal blood glucose concentrations can be maintained. On the other hand, it is possible that the hyperinsulinemia is a primary abnormality. In this case, the high insulin concentrations could cause hypoglycemia, which in turn could be prevented by the development of resistance to the action of insulin on glucose utilization. Despite resistance to insulin action on glucose utilization, fat metabolism (measured by inhibition of fat breakdown) remains sensitive to insulin, even in Pimas with severe diabetes (Howard et al., 1979).

Considering these findings on insulin secretion and action, we hypothesized (Knowler et al., 1982) that the following factors are involved in the development of obesity and diabetes in the Pima Indians. Pima Indians have resistance to glucose utilization and high insulin concentrations in blood, but they appear to have relatively normal sensitivity to insulin actions on fat stores. This could result in increased fat storage when food is available. Thus Neel's "thrifty genotype" may mediate differences in sensitivity to insulin in various metabolic pathways. In times of alternating feast and famine, these differences in insulin sensitivity may have survival value by increasing fat storage. When food supplies are steady and abundant, as they are for the Pimas today, increased fat storage is detrimental, leading to increased obesity, insulin resistance, and eventually diabetes.

ACKNOWLEDGMENTS

We thank the Gila River Indian Community for its cooperation and participation, the staff of the NIADDK for conducting the examinations, and Drs. B.V. Howard and M. Nagulesparan for discussions of insulin resistance.

LITERATURE CITED

Barnett, AH, Eff, C, Leslie, RDG, Pyke, DA (1981) Diabetes in identical twins. Diabetologia *20:*87–93.

Bennett, PH, Rushforth, NB, Miller, M, LeCompte, PM (1976) Epidemiologic studies of diabetes in the Pima Indians. Recent Prog. Horm. Res. *32:*333–375.

Briggs, BR, Jackson, WPU, DuToit, ED, and Botha, MC (1980) The histocompatibility (HLA) antigen distribution in diabetes in Southern African Blacks (Xhosa). Diabetes *29:*68–71.

Cohen, EM (1954) Diabetes mellitus among Indians of the American southwest: its prevalence and clinical characteristics in a hospitalized population. Ann. Intern. Med. *40:*588–599.

Cohen, AM (1961) Prevalence of diabetes among different ethnic Jewish groups in Israel. Metabolism *10:*50–58.

Haury, EM (1976) The Hohokam: Desert farmers and craftsmen, excavations at Snaketown, 1964–1965. Tucson: University of Arizona Press.

Howard, BV, Savage, PJ, Nagulesparan, M, Bennion, LJ, Unger, RH, and Bennett, PH (1979) Evidence for marked sensitivity to the antilipolytic action of insulin in obese maturity-onset diabetics. Metabolism *28:*744–750.

Hrdlička, A (1908) Physiological and Medical Observations Among the Indians of Southwestern United States and Northern Mexico. Washington, DC: Smithsonian Institute, Bureau of American Ethnology, Bulletin 24.

Irvine, WJ, Toft, AD, Holton, DE, Prescott, RJ, Clarke, BF, and Duncan, LJP (1977) Familial studies of type-I and type-II idiopathic diabetes mellitus. Lancet *ii:*325–328.

Joslin, EP (1940) The university of diabetes: A survey of diabetes morbidity in Arizona. J. Am. Med. Assoc. *115:*2033–2038.

Kawate, R, Yamakido, M, Nishimoto, Y, Bennet, PH, Hamman, RF, and Knowler, WC (1979) Diabetes mellitus and its vascular complications in Japanese migrants on the Island of Hawaii. Diabetes Care *2:*161–170.

Khosla, T, and Lowe, CR (1967) Indices of obesity derived from body weight and height. Br. J. Prev. Med. *21:*122–128.

Knowler, WC, Bennett, PH, Hamman, RF, Miller, M (1978) Diabetes incidence and prevalence in Pima Indians: A 19-fold greater incidence than in Rochester, Minnesota. Am. J. Epidemiol. *108:*497–505.

Knowler, WC, Bennett, PH, Bottazzo, GF, Doniach, D (1979) Islet cell antibodies and diabetes mellitus in Pima Indians. Diabetologia *17:*161–164.

Knowler, WC, Pettitt, DJ, Savage, PJ, Bennett, PH (1981) Diabetes incidence in Pima Indians: Contributions of obesity and parental diabetes. Am. J. Epidemiol. *113:*144–156.

Knowler, WC, Savage, PJ, Nagulesparan, M, Howard, BV, Pettitt, DJ, Lisse, JR, Aronoff, SL, Bennett, PH (1982) Obesity, insulin resistance, and diabetes mellitus in the Pima Indians. In J Köbberling and RB Tattersall (eds): The Genetics of Diabetes Mellitus, Serono Symposium No. 47. London: Academic Press, pp. 243–250.

Nagulesparan, M, Savage, PJ, Unger, RH, Bennett, PH (1979) A simplified method using somatostatin to assess in vivo insulin resistance over a range of obesity. Diabetes *28:*980–983.

Nagulesparan, M, Savage, PJ, Knowler, WC, Johnson, GC, Bennett, PH (1982) Increased in vivo insulin resistance in nondiabetic Pima Indians compared with Caucasians. Diabetes *31:*952–956.

National Diabetes Data Group (1979) Classification of diabetes mellitus and other categories of glucose intolerance. Diabetes *28:*1039–1057.

Neel, JV (1962) Diabetes mellitus: A "thrifty genotype" rendered detrimental by "progress"? Am. J. Hum. Genet. *14:*353–362.

Neel, JV (1982) The thrifty genotype revisited. In J Köbberling and RB Tattersall (eds): The Genetics of Diabetes Mellitus, Serono Symposium No. 47. London: Academic Press.

Palumbo, PJ, Elveback, LR, Chu, C-P, Connolly, DC, Kurland, LT (1976) Diabetes mellitus: Incidence, prevalence, survivorship, and causes of death in Rochester, Minnesota, 1945–1970. Diabetes *25:*566–573.

Parks, JH, and Waskow, E (1961) Diabetes among the Pima Indians of Arizona. Ariz. Med. *18:*99–106.

Pettitt, DJ, Baird, HR, Aleck, KA, Lisse, JR, Knowler, WC (1981) Obesity in children following maternal diabetes mellitus during gestation. Am. J. Epidemiol. *114:*437.

Pettitt, DJ, Baird, HR, Aleck, KA, Knowler, WC (1982) Diabetes mellitus in children following maternal diabetes mellitus during gestation. Diabetes *31:*66a.

Pettitt, DJ, Baird, HR, Aleck, KA, Bennett, PH, Knowler, WC (1983) Excessive obesity in offspring of Pima Indian women with diabetes during pregnancy. N. Engl. J. Med. *308:*242–245.

Prior, AM, Beaglehole, R, Davidson, F, Salmond, CE (1978) The relationship of diabetes, blood lipids, and uric acid levels in Polynesians. Adv. Metab. Disord. *9:*241–260.

Reaven, GM, Bernstein, R, Davis, B, Olefsky, JM (1976) Nonketotic diabetes mellitus: Insulin deficiency or insulin resistance? Am. J. Med. *60:*80–88.

Russell, F (1908) The Pima Indians. Twenty-sixth Annual Report of the Bureau of American Ethnology, Washington, DC.

Sarjeantson, SW, Ryan, DP, Ram, P, Zimmet, P (1981) HLA and non-insulin dependent diabetes in Fiji Indians. Med. J. Aust. *1:*462–463.

West, KM (1978) Epidemiology of Diabetes and its Vascular Lesions. New York: Elsevier North Holland.

Williams, RC, Knowler, WC, Butler, WJ, Pettitt, DJ, Lisse, JR, Bennett, PH, Mann, DL, Johnson, AH, Terasaki, PI (1981) HLA-A2 and type 2 (insulin independent) diabetes mellitus in Pima Indians: An association of allele frequency with age. Diabetologia *21:*460–463.

Zimmet, P (1979) Epidemiology of diabetes and its macrovascular manifestations in Pacific populations: The medical effects of social progress. Diabetes Care *2:*144–153.

AMERICAN JOURNAL OF PHYSICAL ANTHROPOLOGY 62:115–118 (1983)

Cerumen Phenotype and Epithelial Dysplasia in Nipple Aspirates of Breast Fluid

NICHOLAS L. PETRAKIS
Department of Epidemiology and International Health, University of California, San Francisco, California 94143

KEY WORDS Cerumen phenotype, Epithelial dysplasia, Nipple aspirates, Breast fluid

ABSTRACT The relationship between presence of dysplastic epithelial cells in nipple aspirates of breast fluid and wet or dry cerumen phenotype was studied in 1,150 white and Asian American women. A statistically significant greater proportion of premenopausal white women of wet cerumen phenotype, compared to women of the dry cerumen type, was found to have cytologic dysplasia (relative risk 6.5 [1.8–22.3]). The effect was not observed in postmenopausal women. The finding offers new support for our hypothesis that an apocrine genetic factor affecting breast gland secretion may influence exposure of the breast epithelium to potential carcinogenic substances of exogenous or endogenous origin.

The female breast is a specialized part of the apocrine gland system, which includes the apocrine sweat glands and the cerumenous (ear wax) glands. We and other investigators have found cerumen phenotype to be a useful genetic marker of apocrine system function (Matsunaga, 1962; Petrakis, 1969, 1971). Cerumen phenotype is inherited as a simple Mendelian trait in which the wet allele is dominant over the homozygous recessive dry allele and is manifested as wet and dry ear wax, respectively. Using a nipple aspiration technique, we found striking differences in breast fluid secretion between women of the wet and women of the dry cerumen phenotype. The highest proportion of secretors occurred among Caucasian women of the wet cerumen type and the lowest proportion among Asian women of the dry cerumen type (Petrakis et al., 1975, 1981a). Also, women with wet cerumen yielded more fluid by nipple aspiration than women with dry cerumen. Thus, the alleles that determine cerumen type and apocrine sweat gland function are associated with the metabolic and secretory activity of the breast.

Biochemical studies of nipple aspirates of breast secretion indicate that a variety of chemical substances of exogenous and endogenous origin are secreted into breast fluid, including putative mutagens and hormones (Petrakis et al., 1978, 1980, 1981b; Wynder and Hill, 1977). Such substances may induce biochemical and morphological changes in breast epithelium. The lower breast cancer rate reported in Asian women may be related to lower metabolic activity in the epithelium of the nonlactating breast (Petrakis, 1977). Low metabolic and secretory activity of breast epithelium would result in lower exposure of the epithelium to harmful substances and protect it from such stimuli. Conversely, metabolically active epithelia would experience higher exposure. New data presented here support an association between wet cerumen phenotype and dysplasia of exfoliated epithelial cells in nipple aspirates and are in agreement with the stated hypothesis.

MATERIALS AND METHODS

Nipple aspiration was performed with a modified Sartorius breast pump and the technique previously described (Petrakis et al., 1975; Sartorius et al., 1977). Cytologic analysis of epithelial cells in breast fluid was performed by the method of King et al. (1975a,b, 1980). Epithelial cells were classified as normal, benign hyperplasia, or atypical hyperplasia/dysplasia. Atypical hyperplastic and dysplastic epithelial cells in nipple aspirates occur in groups and single cells are of varying nuclear size and have altered chromatin pattern. Atypical cell changes ranging from mild to severe

Received August 5, 1982; accepted March 16, 1983.

were designated as "dysplasia." Wet cerumen was readily distinguished from dry by otoscopic examination employing the criteria described previously (Matsunaga, 1962; Petrakis, 1969, 1971). Cerumen type and cytologic interpretation were independently determined.

Breast fluid and cerumen were obtained from 1,150 volunteer women examined at the Breast Screening Clinic at the University of California, San Francisco, who were self- or physician-referred for a variety of breast complaints; none had breast cancer; 1,099 were white and 51 were Asian. Among white women, 697 were premenopausal, 402 had undergone natural or artificial menopause; 51 premenopausal and 43 postmenopausal women had dry cerumen. The 51 Asian women were all premenopausal, 21 of whom had dry cerumen.

The association of cytologic dysplasia with wet and dry cerumen phenotype was determined by means of the risk ratio and its 95% confidence interval. The summary relative risk, controlling for age, was estimated by the method of Mantel-Haenszel (1959), and its confidence limits were obtained by the method of Cornfield, as described by Gart (1970).

RESULTS

Table 1 shows the proportions of premenopausal and postmenopausal white women who had dysplastic epithelial cells in breast fluid, by age and menstrual status, for wet and dry cerumen types. Among premenopausal white women, 18% had epithelial dysplasia, and of these, a significantly higher proportion with the wet cerumen phenotype had epithelial dyslasia compared to the dry cerumen phenotype (20.1% vs. 3.9%, respectively) (Table 2). Adjusting for age, the relative risk of dysplasia associated with wet cerumen was 6.5 (1.7–24.0). No difference in the occurrence of dysplasia was found between postmenopausal women with wet and dry cerumen, even when type of menopause was taken into account.

A lower proportion of the 51 premenopausal Asian women had dysplasia compared to white women (9.8% vs. 18.0%). The Asian women showed a trend similar to that found in white women in the occurrence of dysplasia between

TABLE 1. Proportion of white women with epithelial dysplasia in breast fluid according to age, cerumen phenotype, and menstrual status

		Premenopausal		Postmenopausal	
Age years	Cerumen pheno-type	N dysplasia/ total	% dysplasia	N dysplasia/ total	% dysplasia
<30	Wet	15/104	14.4	—	—
	Dry	0/5	—	—	—
31–40	Wet	55/258	21.3	2/21	9.5
	Dry	1/19	5.3	0/2	—
41–50	Wet	55/262	21.0	16/124	12.9
	Dry	1/25	4.0	3/16	18.8
51–60	Wet	5/22	22.7	32/168	19.0
	Dry	0/2	—	3/16	18.8
>61	Wet	—	—	8/46	17.4
	Dry	—	—	2/9	22.2

TABLE 2. Relative risk of epithelial dysplasia in white women according to cerumen phenotype and menstrual status

	Cerumen pheno-type	N dysplasia/ total	%	Relative risk[1]	95% confidence interval
Premenopausal	Wet	130/646	20.1	6.5*	1.7–24.0
	Dry	2/51	3.9	1.0	—
Postmenopausal	Wet	58/359	16.2	0.9	0.4–1.9
	Dry	8/43	18.6	1.0	—

[1]Mantel-Haenszel estimates of relative risk, adjusted for age. In all comparisons the referent category was the dry cerumen.
*P = .01–.001.

wet and dry cerumen phenotypes: wet 4/30 (13.3%); dry 1/21 (4.8%). However, the sample of Asian women was small, and the elevated relative risk was not statistically significant (3.08 [0.34–27.67]).

DISCUSSION

Several recent studies have shown that the risk of breast cancer is increased in women with benign breast disease if their breast biopsies show atypical or dysplastic epithelial proliferation (Black et al., 1972; Wellings et al., 1975; Kodlin et al., 1977). Such histologic changes in breast glands should be detectable in exfoliated epithelial cells in nipple aspirates of breast fluid. Studies by King and associates (King et al., 1975a, b, 1980, 1981) have found dysplasia of epithelial cells in nipple aspirates to be strongly associated with a histopathological diagnosis of proliferative benign breast disease and of breast cancer, demonstrating the ability of nipple aspiration to detect atypical proliferative breast disease.

Based on our studies of secretion of exogenous chemical substances by the breast glands, we hypothesized that women with genetically determined dry cerumen, compared to women with wet cerumen, have lower metabolic and secretory activity of the breast epithelium and thus lower exposure to potentially damaging substances Petrakis, 1977). Our present results are in agreement with the hypothesis—i.e., epithelial dysplasia in nipple aspirates occurs significantly more frequently in premenopausal white women with wet cerumen than in those with dry cerumen. A similar, although not statistically significant, increased frequency of dysplasia was found in premenopausal Asian women with wet cerumen.

Several caveats should be considered in connection with the findings. The number of women with dry cerumen from whom breast fluid can be obtained is small. Of necessity, the data pertain only to women in whom breast fluid was aspirated. Breast fluid can be obtained from approximately 70% Caucasian and 30% Asian premenopausal women; in postmenopausal women, the proportion of secretors falls to about 30% in Caucasians and 15% in Asians. Although it is not certain that the cytologic results would be similar in women from whom breast fluid cannot be obtained by nipple aspiration, in other studies we found that the small proportion of women with dry cerumen from whom breast fluid can be aspirated yielded significantly less fluid than women with wet cerumen (Petrakis et al., 1981a). Thus, breast glandular secretory activity as estimated by nipple aspiration is lower in women of the dry phenotype, which is consistent with an earlier observation that scanty axillary apocrine secretion is associated with dry cerumen (Yamashita, 1939). Physiological and biochemical studies of the rate of turnover of breast secretion in women are lacking. However, clinical evidence for our contention that the breast glands are metabolically less active in women with the dry apocrine phenotype is based on the very low proportion of secretors among these women and the smaller quantity of fluid, compared with the wet phenotype, which can be aspirated in those few who do secrete fluid. Due to the small numbers of women with dry cerumen it was not possible to assess the possible influence of other breast cancer risk factors.

We have no clear explanation for the absence of a relationship between dysplasia and cerumen phenotype in postmenopausal women. Most likely it is related to the marked decrease in breast secretion ocurring postmenopausally, especially marked among women with dry-type cerumen. We have noted that a higher proportion of postmenopausal women with clinically defined benign breast disease tend to secrete fluid compared to women with clinically normal breasts (Petrakis et al., 1981a). Perhaps a greater proportion of the postmenopausal women with "dry-type" cerumen compared to women with "wet-type" cerumen produce fluid that contains exfoliated dysplastic cells as a result of subclinical breast disease. Also, the results in postmenopausal women are based on a very small number of women with dry-type cerumen who produced sufficient fluid for sampling and cytologic examination.

It is interesting to recall that the wet- and dry-type apocrine genes show marked differences in distribution among populations throughout the world and correspond to differing rates of breast cancer and benign breast disease in these populations. Asian women with a high frequency of dry-type cerumen have low breast cancer risk whereas western women having high frequencies of wet cerumen have high breast cancer rates (Petrakis, 1971). It is reasonable to suggest that the regional differences in potential carcinogenic or toxic environmental factors interacting with wet and dry apocrine genetic factors affecting breast secretion may lead to differences in their effects on breast epithelium. The uptake and metabolism of potentially harmful substances by the breast

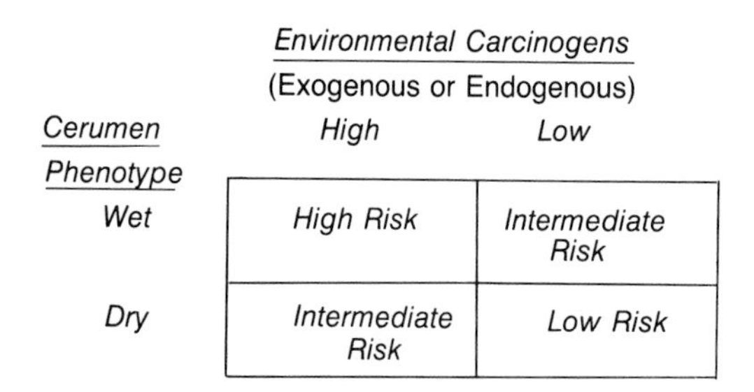

Fig. 1. Model of interaction of cerumen phenotype with differing levels of exposure to environmental carcinogens and predictecd risk of dysplasia of breast epithelium.

epithelium and the development of dysplasia would differ in women who live in high or low cancer-promoting environments (Fig. 1).

The present findings demonstrate an association between cerumen type and the presence of dysplastic epithelial cells indicative of severe breast disease.

ACKNOWLEDGMENTS

I thank Eileen B. King, M.D., Mary-Claire King, Ph.D., Diana Barrett, C.T. (ASCP), Lois Kromhout, and Karen Chew, from the University of California, San Francisco, for their valuable suggestions and technical assistance.

This work was supported in part by USPHS grant PO1-CA-13556-11 from the National Cancer Institute, Bethesda, Maryland.

LITERATURE CITED

Black, MM, Barclay, THC, Cutler, SJ, Hankey, BF, and Asire, AJ (1972) Association of atypical characteristics of benign breast lesions with subsequent risk of breast cancer. Cancer *29:*338–343.

Gart, JJ (1970) Point and interval estimation of the common odds ratio in the combination of 2 × 2 tables with fixed marginals. Biometrika *57:*471–475.

King, EB, Barrett, D, King M-C, and Petrakis, NL (1975a) Cellular composition of the nipple aspirate specimen of breast fluid. I. The benign cells. Am. J. Clin. Pathol. *64:*728–738.

King, EB, Barrett, D, and Petrakis, NL (1975b) Cellular composition of the nipple aspirate specimen of breast fluid. II. Abnormal findings. Am. J. Clin. Pathol. *64:*739–748.

King, EB, Zimmerman, AL, Barrett, DL, Petrakis, NL, and King, M-C (1980) Cytopathology of abnormal mammary duct epithelium. In HE Nieburgs (ed): Cancer Detection in Specific Sites, Vol. 2. New York: Marcel Dekker, pp. 1831–1845.

King, EB, Chew, K, Ernster, VL, Barrett, DL, and Petrakis, NL (1981) Abnormal nipple aspirate cytology and proliferative changes associated with malignant breast disease. Presented, 29th Annu. Meeting Am. Soc. Cytol. Manuscript submitted for publication.

Kodlin, D, Winger, EE, Morgenstern, NL, and Chen, U (1977) Chronic mastopathy and breast cancer. A follow-up study. Cancer *39:*2603–2607.

Mantel, N, and Haenszel, W (1959) Statistical aspects of the analysis of data from retrospective studies of disease. JNCI *22:*719–748.

Matsunaga, E (1962) The dimorphism in human normal cerumen. Ann. Hum. Genet. *25:*273–286.

Miettinen, OS (1976) Estimability and estimation. Am. J. Epidemiol. *103:*226–235.

Petrakis, NL (1969) Dry cerumen—a prevalent genetic trait among American Indians. Nature *222:*1080–1081.

Petrakis, NL (1971) Cerumen genetics and human breast cancer. Science *173:*347–349.

Petrakis, NL, Mason, L, Lee, RE, Sugimoto, B, Pawson, S, and Catchpool, F (1975) Association of race, age, menopausal status, and cerumen type with breast fluid secretion in nonlactating women as determined by nipple aspiration. JNCI *54:*829–834.

Petrakis, NL (1977) Breast secretory activity in nonlactating women, postpartum breast involution and the epidemiology of breast cancer. Natl. Cancer Inst. Monogr. *47:*161–164.

Petrakis, NL, Gruenke, LD, Beelen, TC, Craig, JC, and Castagnoli, N Jr (1978) Niccotine in breast fluid of nonlactating women. Science *199:*303–305.

Petrakis, NL, Maack, CA, Lee, RE, and Lyon, M (1980) Mutagenic activity of nipple aspirates of breast fluid. Cancer Res. *40:*188–189.

Petrakis, NL, Ernster, VL, Sacks, ST, King, EB, Schweitzer, RJ, Hunt, TK, and King, M-C (1981a) Epidemiology of breast fluid secretion: Association with breast cancer risk factors and cerumen type. JNCI *67:*277–284.

Petrakis, NL, Gruenke, LD, Craig, JC (1981b) Cholesterol and cholesterol epoxides in nipple aspirates of human breast fluid. Cancer Res. *41:*2563–2565.

Sartorius, OW, Smith, HS, Morris, P, Benedict, D, and Fiesen, L (1977) Cytologic evaluation of breast fluid in the detection of breast disease. JNCI *59:*1073–1080.

Wellings, SR, Jensen, HM, and Marcum, RG (1975) An atlas of subgross pathology of the human breast with special reference to possible precancerous lesions. JNCI *55:*231–273.

Wynder, EL, and Hill, P (1977) Prolactin, estrogen, and lipids in breast fluid. Lancet *2:*840–841.

Yamashita, S (1939) On the offensive odor of the armpit and the soft earwax among Formosans in Canton. Jpn. J. Anthropol. *54:*444–446.

AMERICAN JOURNAL OF PHYSICAL ANTHROPOLOGY 62:119–126 (1983)

Genetic and Anthropologic Factors in Gluten-Sensitive Enteropathy

WARREN STROBER
Immunophysiology Section, Metabolism Branch, National Cancer Institute, National Institute of Health, Bethesda, Maryland 20014

KEY WORDS Gluten-sensitivity, Histocompatibility antigens, Linkage disequilibrium, Immune response genes, Autoimmunity.

ABSTRACT Gluten-sensitive enteropathy (GSE), a disease characterized by intestinal villous atrophy and malabsorption, is due to a sensitivity to wheat protein, gliadin, and probably has its basis in an underlying immunologic defect. GSE has a genetic basis in that some 70–80% of north American and northern European patients bear the HLA-B8 histocompatibility type and about 90% bear the HLA-DRw3 histocompatibility type. These histocompatibility types are both increased because they are in linkage disequilibrium in normal populations. This suggests that HLA-B8 and HLA-DRw3 genes are in linkage disequilibrium with a GSE disease gene, accounting for the association of the disease with certain histocompatibility antigens.

A gene coding for a lymphoid surface antigen has been identified which is not HLA-linked. This gene is distributed at a low frequency in the general population; it has been proposed that both the HLA and non-HLA genes important to GSE code for different domains of the single surface receptor molecule that somehow predisposes to a heightened immune reaction to gliadin, thus causing the disease.

GSE is most prevalent where HLA-B8 occurs at the highest frequency in the general population and is not seen in populations where HLA-B8 is not found. One explanation for this is that the gene complex containing HLA-B8 (and HLA-DRw3) evolved in response to infectious agents: Individuals bearing this complex were capable of more vigorous antibody response. However, such individuals were also more likely to be hypersensitive to wheat protein; as wheat became domesticated these individuals may have been at a disadvantage. It is only in regions such as northern Europe where wheat domestication occurred relatively late that one finds both a high frequency of HLA-B8 and a high incidence of GSE.

Gluten-sensitive enteropathy (GSE) is a disease of small intestine mucosa which is due to wheat (gluten protein/gliadin) toxicity. The latter is manifested as villous flattening (atrophy) which leads to diarrhea and malabsorption of ingested fat, carbohydrate, and protein. Affected persons experience weight loss as well as a spectrum of symptoms related to malabsorption.

The role of genetic factors in gluten-sensitive enteropathy is best understood within the context of current hypotheses concerning the pathogenesis of the disease (Strober, 1976, 1978). It is now believed that in GSE the ingestion of gluten protein results in the induction (within the organized gut lymphoid tissue, the Peyer's patches) of immunocompetent cells with specificity for gluten. These cells include B cells capable of differentiating into antibody-producing plasma cells as well as T cells capable of mediating cytolysis of appropriate target cells. It is presumed that such cells ultimately migrate to the lamina propria where they (or their products) interact with gluten protein bound to epithelial cell targets and thereby cause epithelial cell injury.

The induction of gluten protein-specific immune effector cells which mediate tissue damage cannot be considered normal for the gastrointestinal tract. On the contrary, we know

Received August 5, 1982; accepted March 16, 1983.

that when a food protein or any orally ingested antigen is introduced into the gastrointestinal lumen, nonresponsiveness or tolerance to the oral antigen is a frequent result (Strober, 1981). In some instances this has been associated with the appearance of suppressor T cells, suppressor B cells, or suppressor factors which are specific for the oral antigen (Asherson et al., 1977; Kagnoff, 1978; Richman et al, 1978). On this basis the immune response to gluten protein in patients with GSE can be considered an aberration whose cause is likely to be central to the initiation of disease. It is at this point that genetic mechanisms probably play a key role, particularly since a major genetic factor found in GSE, a gene in the major histocompatibility complex, can quite conceivably be associated with genes that participate in the control of immune responses (Benacerraf and McDevitt, 1972).

HLA ANTIGENS ASSOCIATED WITH GSE

As soon as GSE became firmly established as a diagnostic entity, it became clear that genetic factors play a role in disease pathogenesis. This was first shown in numerous family studies of the disease in which it was observed that about 10% of individuals with GSE also had an affected sibling, and a somewhat lower fraction had a parent with the disorder (MacDonald et al., 1965; Mylotte et al, 1974; Shipman et al., 1975; Stokes et al., 1976). From these family studies, however, it could not be established if the disease was due to a single gene (dominant or recessive) or was the result of several genes acting in concert. In the early 1970s a major advance occurred when GSE was found to be associated with the histocompatibility antigen HLA-B8. In initial studies, it was found that about 80% of GSE patients, but only 20% of normal individuals, carried the HLA-B8 antigen (Falchuk et al., 1972; Stokes et al., 1972). In follow-up studies, based on the HLA typing of hundreds of patients with GSE, it was calculated that the chance of having GSE was eight to nine times greater if one were HLA-B8 positive than if one were HLA-B8 negative (relative risk factor) (Strober, 1977). These subsequent studies also indicated that there was a correlation between the prevalence of HLA-B8 in GSE and the proportion of individuals in the general population bearing HLA-B8: in those populations exhibiting a relatively high frequency of HLA-B8 (populations in the U.S. and in northern Europe) there was a 75–85% prevalence of HLA-B8 in GSE, whereas in those populations exhibiting a low frequency of HLA-B8 (in southern Europe) the prevalence of HLA-B8 in GSE was around 45%. Interestingly, in Oriental populations where HLA-B8 is not found, GSE is also not found (Bodmer and Thomson, 1977).

GSE is also associated with other HLA abnormalities. Initially, it was found that the HLA-D locus antigens, HLA-Dw3 (defined by HLA-typing cells), and the related antigen, HLA-DRw3 (defined by HLA-typing sera), were increased in GSE. The percentage of GSE patients with Dw3/DRw3 probably exceeds the percentage with HLA-B8 and is in the 85–95% range; this contrasts with the 20–25% range found in normal individuals (Keuning et al., 1976). Later, another HLA-D locus antigen, DRw7, was found in 60–70% of Italian and Spanish GSE patients but in only 25–40% of control patients (Betuel et al., 1980). Inexplicably, this antigen was not higher in frequency among northern European or U.S. patient populations.

Finally, studies of HLA antigen distribution in families of GSE patients reveal several important facts. First, within families, GSE segregates with HLA-B8 and HLA DRw3 (Harms et al., 1974). Second, in families with more than one affected member, the diseased individuals are at least partially HLA identical and are never totally HLA nonidentical (Falchuk et al., 1978). Third, patients with GSE quite frequently have HLA-identical siblings who are nevertheless disease-free. The last fact is important because it implies that HLA genes are not the only factor necessary for disease, as will be discussed below.

The association of both HLA-B8 and HLA-DRw3 genes (as well as the possible association of other HLA genes) with GSE is not unexpected, as these HLA genes are in linkage disequilibrium, i.e., they tend to occur together in the normal population. Such nonrandom pairing (or grouping) of genes is also seen with other HLA gene combinations and may occur because the genes involved participate in a complementary function which is advantageous to the host. Linkage disequilibrium can also explain the fact that both HLA-B8 and HLA-DRw3 are associated with disease yet are present in many normal individuals who do not have disease. This is so if one assumes that the increased occurrence of these HLA genes in GSE is due to their linkage disequilibrium with another gene, an as yet unidentified GSE "disease gene." In this view, the HLA-B8 and HLA-DRw3 genes are not directly involved in disease pathogenesis but are merely bystander or marker genes. However, there is now evidence that genes not in the HLA locus must be present for the development of GSE (see below). Thus, one does not have to invoke an unidentified disease gene in the HLA locus as necessary for GSE to explain the fact that HLA-B8 and HLA-DRw3 are found in normal individuals as well as patients with disease. Further, the existence of such non-HLA genes suggests that HLA-B8 and HLA-DRw3 are directly involved in disease pathogenesis.

POSSIBLE MODES OF HLA GENE FUNCTION IN GSE

Two general possibilities can be put forward to explain the mechanism of HLA genes in the pathogenesis of GSE. On the one hand, these genes could act in a nonspecific fashion that places individuals at risk for GSE as well as several other diseases. In this view, HLA genes always act in conjunction with other genetic or environmental factors to bring about a particular disease state. Alternatively, these genes could act in a specific fashion which results in a unique pathophysiologic mechanism leading inevitably to GSE and only GSE.

Turning first to the possibility of a nonspecific mechanism, it is now known that HLA-B8/DRw3 is associated with increased immune responses to several different antigens in populations of normal individuals. This suggests that HLA-B8/DRw3 genes code for background immune hyperresponsiveness which, in concert with genes controlling response to specific antigens, leads to disease.

Several studies support the occurrence of background immune hyperresponsiveness in normal HLA-B8-positive (or HLA-DRw3-positive) individuals. Cunningham-Rundles et al. (1978) have shown that cells obtained from normal individuals bearing HLA-B8 have increased in vitro proliferative responses to gluten protein as compared to individuals not bearing HLA-B8. Osoba and Falk (1978) have shown that the magnitude of mixed lymphocyte responses (proliferative response to T cells when mixed with allogeneic B cells and macrophages) was greater in individuals bearing HLA-B8 than in individuals bearing any other HLA type. Finally, Griffing et al. (1980) have observed that HLA-DRw3 is elevated in a patient population characterized by elevated anti-DNA titers, regardless of their underlying disease; in addition, these authors found that HLA-DRw3 was elevated only in those patients with systemic lupus erythematosus (SLE) who had anti-DNA, thereby relating this histocompatibility type to the autoantibody rather than to the disease.

These studies of the relation between HLA-B8/DRw3 and immune responses in normal individuals should not be construed as indicating that HLA-B8/DRw3 is associated with a global hyperresponsiveness to all antigens. On the contrary, Cunningham-Rundles et al. (1978) found that responses to candida antigen did not differ in HLA-B8-positive and -negative individuals. Similarly, studies of responses to a variety of proteins associated with infectious agents such as influenza and vaccinia vaccines, as well as streptococcal proteins and schistosomal worm antigens, were not greater in HLA-B8/DRw3-positive individuals, although they were increased or decreased in individuals with several other HLA types (depending on the particular antigen studied) (DeVries et al., 1977; Greenberg et al., 1980; Sasazuki et al., 1980; Spencer et al., 1976).

HLA-B8/DRw3 is also associated with a variety of other disease states (though to a less striking degree), thus adding further support to a nonspecific mechanism. This disease states are unlike GSE and unlike each other, but their shared feature is that each is an abnormality of immune function and each is characterized by the production of antibodies and/or other immune effectors with specificity for self-antigens or modified self-antigens. In addition, these disease states are frequently characterized by decreased nonspecific suppressor T cell function which, in turn, is responsible for increased immune responses to a variety of stimuli (Sakane et al., 1978; Zilko et al., 1979).

One may conclude, therefore, that HLA-B8/DRw3 genes are associated with abnormalities of the immune response in which excessive immune responses are a characteristic feature.

The HLA-B8/DRw3 genes do not occur in all autoimmune patients. Thus, when evaluating the significance of their association with diseases of autoimmunity, the conclusion that these genes are necessary or sufficient for the underlying immunologic abnormality is not warranted. Specifically, while suppressor T cell defects have been found in SLE and myasthenia gravis, diseases associated with an increased frequency of HLA-B8/DRw3, there exists at least one autoimmune state, primary biliary cirrhosis, where decreased suppressor cell function can be quite convincingly demonstrated, yet the prevalence of HLA-B8 among patients with this disease is normal (James et al., 1980). Further, in GSE, where HLA-B8/DRw3 is most prominently increased, no evidence yet exists for or against a suppressor defect.

With respect to the concept that the HLA genes associated with GSE are active in a specific fashion that is unique to GSE, it may be that the HLA gene or genes involved are immune response genes coding for hypersensitivity to gluten protein. Immune T cells which regulate immune responses become activated by exogenous (foreign) antigens only when they impinge upon such cells in concert with a particular class of histocompatibility gene products, i.e., I region gene products in the mouse and D region gene products in man. The actual activation process may involve simultaneous but separate signals from both foreign antigen and histocompatibility antigen (dual specificity) or may involve a combined signal composed of these antigens jointly (altered self specificity).

Since I(D) region gene products are intimately involved with the process of antigen recognition, it is clear in principle how HLA antigens can influence particular immune responses. The details of the process, however, are less clear. It is possible that molecules coded for by I(D) region genes determine if and how antigens will be presented to T cells and therefore that nonresponsiveness and responsiveness are related to the ability of antigen-presenting cells to bind and/or orient antigens in an immunogenic or nonimmunogenic manner. One of many studies supporting this concept showed that cloned T cells (i.e., T cells arising from a single progenitor cell) respond differently to antigen depending on whether the antigen is presented on macrophages obtained from genetically high or low responder animals (Heber-Katz et al., 1982).

On the other hand, it is possible that I(D) region genes determine immune responses by virtue of their involvement in shaping the repertoire of T cells which develops by somatic selection in the thymus. Such involvement could take several forms, but in each case, the end result is the creation of a T cell repertoire which can or cannot respond to a particular array of foreign antigen plus I(D) region antigen. This view is supported by the observation that T cells from one nonresponder strain (a strain of animal which gives a low response to a given antigen) can respond to the antigen if it is presented on an antigen-presenting cell from an animal of another strain, even if the antigen-presenting cell in question is from an animal of a second nonresponder strain (Ishii et al., 1981).

To apply these concepts to the possible role of histocompatibility genes and responses to particular antigens which might lead to disease, one could propose that certain individuals (with certain arrays of histocompatibility antigens) respond to a foreign antigen either because that individual possesses, by virtue of the I(D) region genes present, antigen-presenting cells capable of presenting that antigen in an immunogenic manner, or because that individual possesses T cells which recognize that antigen as immunogenic (i.e., the tendency to form activated T cells of the helper regulatory subset and/or the tendency not to form activated T cells of the suppressor regulatory subset) when it is associated with self I(D) region genes. In the case of gluten sensitivity, these genes could result in abnormally active induction of helper T cells for gluten protein responses or these genes could result in decreased suppressor T cells for gluten protein responses. This latter possibility is particularly relevant to an antigen impinging on the immune system via the gastrointestinal tract since, as noted above, such antigens quite often evoke suppressor rather than helper influences.

No direct evidence in favor of the existence of unique histocompatibility genes (immune response genes) dedicated to gluten protein responses is available. In animal systems such genes are identified with the use of chemically simple antigens and inbred mouse strains. By analogy, gluten protein-specific immune response genes could be identified in GSE by studying responses to a defined fraction of gluten protein in GSE families containing one or more individuals with GSE. However, such studies depend on the capacity to measure gluten protein-specific immune responses in vitro and this is not yet possible.

THE ROLE OF HLA-B8 IN THE PATHOGENESIS OF GSE

The involvement of A, B, C, and D locus HLA antigens with disease is widely considered to be an indirect involvement which depends on linkage equilibrium between the HLA genes controlling these antigens (conventional HLA antigens) and actual disease genes which are directly responsible for pathologic effects. If true, how does one account for the fact that it is frequently a particular set of HLA genes (i.e., HLA-B8 and HLA-DRw3), which is in linkage disequilibrium with one or more disease genes in a variety of of disease states? A possible answer is that the HLA-B8 and HLA-DRw3 genes (or other conventional HLA genes) are not mere marker genes, but instead provide some element necessary to the disease process.

Supporting this latter view are experiments by Biddison et al. (1980), who induced cytotoxic effector T cells specific for influenza virus by culturing cells in the presence of this virus in humans. As in murine systems, they found that the effector cells killed (lysed) influenza virus-coated target cells; moreover, the extent of lysis was in direct relation to the extent with which effector and target shared HLA antigens. Supporting the direct involvement hypothesis they showed in family studies that cells obtained from certain individuals preferentially lysed virus-coated target cells of one parent rather than the other, even though both target cell sources had an equal HLA similarity with the effector cell source. These studies, therefore, support the notion that certain exogenous antigens (in this case viral antigens), appearing in the context of certain HLA antigens, initiate better responses or present better targets than the same antigens appearing in the context of other HLA antigens, because of inherent structural characteristics of the exogenous antigen-HLA antigen complex.

In the case of unrelated individuals, Biddison et al. (1980) further found that individuals sharing HLA antigens may nevertheless differ in their capacity to recognize and lyse virus-modified target cells exhibiting the shared antigen. This result suggests that particular HLA genes may be linked, in some individuals, to regulatory genes which enable a high level of response to antigens associated on the cell surfaces with the linked HLA antigen. It is possible that certain A and B locus HLA genes are linked to regulatory HLA genes which control responses to antigens associated with the gene product of the A and B locus HLA genes; thus, two genes are assumed to work in tandem to determine the magnitude of the immune response: gene A coding for A and B locus HLA antigens which, in association with exogenous antigen, is involved in immunostimulation, and gene B (linked gene) involved in immunoregulation of the response to exogenous antigen associated with the product of gene A.

These considerations suggest a basis for linkage disequilibrium since they indicate that certain HLA gene pairs code for either decreased or increased immune responses which have positive survival value. For example, individuals bearing certain HLA gene combinations may automatically possess the capacity to respond vigorously to noxious viruses (exogenous antigens) because the associated cell surface HLA antigens with which the viruses are necessarily associated during immune responses are coded for by HLA genes which are linked to regulatory genes that provide for enhanced responses to viruses associated with the HLA antigens. Similarly, the paired genes in linkage disequilibrium could also have a negative survival value in that they could lead to a propensity to respond against self-antigens. This could result from the fact that an enhanced immune response could occur that is specific for self or modified self components which are analogous to the exogenous antigens discussed above.

On the basis of this experimental and theoretical framework, it is reasonable to suggest that the HLA-B8/DRw3 genes are more than bystander or marker genes in autoimmune diseases and in gluten-sensitive enteropathy. Thus, when recognized immunologically in concert with HLA-B8, it is possible that gluten protein forms a strong positive signal or target in the immune response, particularly when HLA-B8 occurs in association with a linked regulatory (Ir) gene represented by HLA-DRw3. This view of the role of the HLA-B8 gene gains some support from the fact that HLA-B8-positive and HLA-B8-negative patients appear to differ in their susceptibility to gluten protein toxicity. Falchuk et al. (1980) assessed HLA-B8-positive and -negative individuals using an in vitro organ culture model of GSE wherein intestinal tissue from patients (and appropriate control individuals) are exposed to gluten protein in vitro. They found that the tissue obtained from HLA-B8-positive patients was more susceptible to in vitro tissue damage than the tissue from HLA-B8-negative individuals. This could occur because of differences in the nature of the immune target formed in HLA-B8-positive and -negative patients, with the former giving rise to more lysable targets than the latter.

NON-HLA-ASSOCIATED GENETIC FACTORS IN GSE

We have seen that the HLA disease gene may be a nonspecific gene effecting immune responses generally or may be a specific gene which causes aberrant responses to specific antigens; in the case of GSE this would be a gene coding for an abnormal gastrointestinal tract response to gluten protein. Whichever of these possibilities ultimately proves true, it remains likely that additional non-HLA genetic factors are necessary for the occurrence of disease. This is immediately obvious in the case of the nonspecific HLA gene hypothesis, but actually is a requirement for the specific HLA hypothesis as well, since analysis of GSE families indicates that HLA-identical siblings are frequently encountered who are nevertheless disease-free.

Supporting the hypothesis of non-HLA genetic factors is the observation that mothers of GSE patients, particularly mothers of young patients, may be sources of antibodies which recognize non-HLA-associated lymphocyte antigens which are associated with GSE (Mann et al., 1976; Peña et al., 1978). These antigens were found on cells obtained from at least 80% of patients whether or not they are HLA-B8 or HLA-DRw3 positive, whereas they are found on less than 10% of normal individuals. The antigens are present on B cells and macrophages but not on T cells, indicating that they differ from A and B locus HLA antigens. Finally, this "GSE-associated B cell antigen" segregates independently of HLA antigens in families; it appears that the maternal antisera thus recognize a non-HLA-associated genetic factor which is controlled by genes not located on chromosome 6 (Peña et al., 1978).

The existence of an identifiable non-HLA gene in GSE leads to certain insights regarding the nature of the genetic control of the disease. First, if the two-gene hypothesis were carried to its logical conclusion, it is apparent GSE is not necessarily associated with a unique HLA disease gene in the sense that only GSE patients bear a particular HLA gene. Rather, it seems possible and even probable that disease is brought about by the simultaneous occurrence of several genes which are each distributed in relatively low frequency in the normal population. Second, knowledge of the individual genetic factors necessary for disease allows one to predict, both in families and in populations, who will have disease. Studies of families containing GSE patients indicate that individuals homozygous for the gene controlling the GSE-associated B cell antigen and who also bear the appropriate HLA antigen (HLA-B8/DRw3) are very likely to have disease. A study applying a two-locus model of GSE inheritance using estimates of HLA and non-HLA disease gene frequencies showed that disease frequencies could be generated that approximated those actually observed, assuming recessive genes at each locus (Greenberg and Rotter, 1981). This study gives rise to the possibility that two genes are both necessary and sufficient to the occurrence of disease.

In summary, the genetic susceptibility to GSE is due to at least two different genes, one located in the major histocompatibility locus and one located elsewhere in the genome. The histocompatibility gene serves either a nonspecific function to induce immunologic hyperresponsiveness or a specific function related to gluten protein. Further studies of GSE and other autoimmune states would differentiate between these possibilities. The nature and function of the second (non-HLA) gene in GSE is presently unknown.

ANTHROPOLOGIC CONSIDERATIONS IN GSE

The genes associated with GSE, HLA-B8, and HLA-DRw3 may have developed linkage disequilibrium during early human evolution because of the positive survival value they conferred in terms of immune response. The possible negative effects that the genes could have had, such as an increased frequency of autoimmunity and/or gluten sensitivity, should have been negligible: autoimmune disease would become a cause of morbidity after child-rearing age, and gluten protein was not part of the human diet. Such adverse effects would become more significant, however, with the advent of wheat domestication. It is therefore reasonable to ask: What impact did the introduction of wheat into the human diet have on the distribution and profile of human histocompatibility HLA genes, on the occurrence of GSE, and, most importantly, on the human species?

We know that early farming originated in the Middle East in conjunction with natural stands of wild grasses containing edible grains (Harlan and Zohary, 1966). Second, there is recent evidence, based on radiocarbon dating of neolithic farming implements in various parts of Europe and Africa that farming spread radially from a central area located in the Middle East (modern-day Israel). The data indicate that farming advanced approximately 1 kilometer/year across the European continent in the westerly direction (Ammerman and Cavalli-Sfoza, 1971). The prevalence of HLA-B8 in the human population is highest in the northwestern reaches of the continent, where farming arrived relatively late (about 5,000 years ago) and is progressively lower as one converges to the area where farming occurred earliest (about 9,000 years ago) (Simoons, 1981). These facts, and the observation that the prevalence of GSE in various parts of Europe is related to the prevalence of HLA-B8, led Simoons (1981) to suggest that this gradient in Europe may be due to negative disease selection pressure on the HLA-B8-positive population. More specifically, exposure to wheat occurred over a longer period of time in southeastern Europe than in northwestern Europe and consequently HLA-B8-positive populations in the former areas were at risk for GSE for a longer period of time. Thus, the residual populations in areas with a relatively longer history of wheat domestication were depleted of HLA-B8-positive individuals as well as individuals capable of developing GSE to a greater extent than the residual population in areas with a relatively shorter period of wheat domestication.

Whether a similar situation exists east of the Middle Eastern cradle of wheat domestication is not known at present. A high HLA-B8 frequency is found in Pakistan and there is some evidence that there is a high prevalence of GSE in this country, making it a geographical and gastric reflection of northwestern Europe as regards HLA-B8 and GSE. The Orient is marked by a lack of both HLA-B8 and GSE: Various HLA genes may have evolved independently, thus never including a gene complex that led to both immune hyperresponsiveness and wheat sensitivity.

The evolutionary impact of wheat sensitivity may be related to the possibility that loss of these genes led to human groups that were less able to cope with certain widespread infections, such as bubonic plague. However, the introduction of wheat would only reduce the number of such individuals, and not eliminate them, as other genetic factors in addition to certain HLA genes are actually necessary for wheat sensitivity. Thus, the effect of the introduction of wheat on susceptibility of human populations to disease might have been relatively marginal.

In conclusion, the HLA-B8 and gluten protein interaction undoubtedly affected human populations, since components of the interaction are important genetic/environmental factors. The nature of the interaction should become clearer as we learn more about the role of the HLA-B8/DRw3 complex in host defense.

LITERATURE CITED

Ammerman, AJ, and Cavalli-Sforza, LL (1971) Measuring the rate of spread of early farming in Europe. Man *6*:674–688.

Asherson, GL, Zembala, M, Perera, MA, Mayhew, B, and Thomas, WR (1977) Production of immunity and unresponsiveness in the mouse by feeding contact sensitizing agents and the role of suppressor cells in the Peyer's patches, mesenteric lymph nodes and other lymphoid tissues. Cell. Immunol. *33*:145–155.

Benacerraf, B, and McDevitt, HO (1972) Histocompatibility-linked immune response genes. Science *175*:273–279.

Betuel, H, Gebuhrer, L, Descos, L, Percebois, H, Minaire, Y, and Bertrand, J (1980) Adult celiac disease associated with HLA-DRw3 and DRw7. Tissue Antigens *15*:231–238.

Biddison, WE, Payne, SM, Shearer, GM and Shaw, S (1980) Human cytotoxic T cells responses to trinitrophenyl hapten and influenza virus. Diversity of restriction antigen and specificity of HLA-linked genetic regulation. J. Exp. Med. *152*:2045–2175.

Bodmer, W, and Thomson, G (1977) Population genetics and evolution of the HLA system. In J Dausset and A Svejgaard (eds): HLA and Disease. Copenhagen: Munksgaard, pp. 280–295.

Cunningham-Rundles, S, Cunningham-Rundles, C, Pollack, MS, Good, RA, and Dupont, B (1978) Response to wheat antigens in *in vitro* lymphocyte transformation among HLA-B8-positive normal donors. Transplant. Proc. *10*:977–979.

DeVries, RRP, Kreeftenberg, HG, Loggen, HG and van Rood, JJ (1977) *In vitro* immune responsiveness to vaccinia virus and HLA. N. Engl. J. Med. *297*:692–696.

Falchuk, ZM, Katz, AJ, Schwachman, J, Rogentine, GN, and Strober, W (1978) Gluten-sensitive enteropathy: Genetic analysis and organ culture study in 35 families. Scand. J. Gastroenterol. *13*:839–843.

Falchuk, ZM, Nelson, DL, Katz, AJ, Bernardin, JE, Kasarda, DD, Hague, NE, and Strober, W (1980) Gluten-sensitive enteropathy. Influence of histocompatibility type on gluten sensitive *in vitro*. J. Clin. Invest. *66*:227–233.

Falchuk, ZM, Rogentine, GN and Strober, W (1972) Predominance of histocompatibility antigen HLA-B8 in patients with gluten-sensitive enteropathy. J. Clin. Invest. *51*:1602–1605.

Greenberg, DA, and Rotter, JI (1981) Investigations of a two locus model for celiac disease. In RB McConnell (ed): The Genetics of Coeliac Disease. Baltimore: M.T.P. Press, pp. 251–262.

Greenberg, LJ, Chopyk, RL, Bradley, PW, and Lalouel, JM (1980) Immunogenetics of response to a purified antigen from group A streptococci, II. Linkage of response to HLA. Immunogenetics *11*:161–167.

Griffing, LW, Moore, SB, Luthra, HS, McKenna, CH, and Fathman, CG (1980) Association of antibodies to native DNA with HLA-DRw3. A possible major histocompatibility complex-linked human immune response gene. J. Exp. Med. *152*:3195–3255.

Harlan, JR, and Zohary, D (1966) Distribution of wild wheats and barley. Science *153*:1074–1980.

Harms, K, Granitsch, G, Rossipal, E, Ludwig, H, Polymenidas, Z, Scholz, S, Wank, R, and Albert, ED (1974) HLA in patients with coeliac disease and their families. In WTJM Hekkens and AS Pena (eds): Coeliac Disease. Leiden: Stenfert Kroese, pp. 215–226.

Heber-Katz, E, Schwartz, RH, Matis, LA, Hannum, C, Fairwell, T, Appella, E, and Hansburg, D (1982) Contribution of antigen-presenting cell major histocompatibility complex gene products to the specificity of antigen-induced T cell activation. J. Exp. Med. *155*:1086–1099.

Ishii, N, Baxevanis, CN, Nagy, ZA, and Klein, J (1981) Responder T cells depleted of alloreactive cells react to antigen presented on allogenic macrophages from nonresponder strains. J. Exp. Med. *154*:978–982.

James, SP, Elson, CO, Jones, EA and Strober, W (1980) Abnormal regulation of immunoglobulin synthesis *in vitro* in primary biliary cirrhosis. Gastroenterology *79*:242–254.

Kagnoff, MF (1978) Effects of antigen-feeding on intestinal and systemic immune responses. III. Antigen-specific serum-mediated suppression of humoral antibody responses after antigen feeding. Cell. Immunol. *40*:186–203.

Keuning, JJ, Pena, AS, van Leeuwen, A, van Hoof, JP, and van Rood, JJ (1976) HLA-Dw3 associated with coeliac disease. Lancet *1*:506–508.

MacDonald, WC, Dobbins, WO and Rubin, CE (1965) Studies of the familial nature of celiac sprue using biopsy of the small intestine. N. Engl. J. Med. *272*:448–456.

Mann, DL, Katz, SI, Nelson, DL, Abelson, L, and Strober, W (1976) Specific B-cell antigens associated with gluten-sensitive enteropathy and dermatitis hepetiformis. Lancet *1*:110–111.

Mylotte, M, Egan-Mitchell, B, Fottrell, PF, McNichol, R, and McCarthy, CF (1974) Family studies in coeliac disease. Q. J. Med. *43*:359–369.

Osoba, D and Falk, J (1978) HLA-B8 phenotype associated with an increased mixed leukocyte reaction. Immunogenetics *6*:425–432.

Peña, AS, Mann, DL, Hague, NE, Heck, JA, van Leeuwen, A, van Rood, JJ, and Strober, W (1978) Genetic basis of gluten-sensitive enteropathy. Gastroenterology *75*:2230–2235.

Richman, LK, Chiller, JM, Brown, WR, Hanson, DG, and Vaz, NH (1978) Enterically induced immunologic tolerance. I. Induction of suppressor T lymphocytes by intragastric administration of soluble proteins. J. Immunol. *121*:2429–2434.

Sakane, T, Steinberg, AD, and Green, I (1978) Studies of immune functions of patients with systemic lupus erythematosus. I. Dysfunction of suppressor T-cell activity related to impaired generation of, rather than response to, suppressor cells. Arthritis Rheum. *21*:657–664.

Sasazuki, T, Ohta, N, Kaneoka, R, and Kojima, S (1980) Association between an HLA haplotype and low responsiveness to schistosomal worm antigen in man. J. Exp. Med. *152*:3145–3185.

Shipman, RT, Williams, AL, Kay, R, and Townley, RRW (1975) A family study of coeliac disease. Aust. N. Z. J. Med. *5*:250–255.

Simoons, FJ (1981) Celiac disease as a geographic problem. In N Kretchmer and D Walcher (eds): Food, Nutrition and Evolution. New York: Masson, pp. 179–199.

Spencer, MJ, Cherry, JD, and Terasaki, PI (1976) HLA antigens and antibody response after influenza A vaccination. Decreased response associated with HLA type W16. N. Engl. J. Med. *294*:13–16.

Stokes, PL, Asquith, P, Holmes, GKT, Mackintosh, P, and Cooke, WT (1972) Histocompatibility antigens associated with adult coeliac disease. Lancet *2*:162–164.

Stokes, PL, Ferguson, R, Holmes, GKT, and Cook, WT (1976) Familial aspects of coeliac disease. Q. J. Med. *45*:567–582.

Strober, W (1976) Gluten sensitive enteropathy. Clin. Gastroenterol. *5*:429–452.

Strober, W (1977) Abnormalities of the HLA system and gastrointestinal disease. In J Dausset and A Svejgaard (eds): HLA and Disease. Copenhagen: Munksgaard, pp. 168–185.

Strober, W (1978) An immunological theory of gluten-sensitive enteropathy. In B McNicholl, CF McCarthy, and PF Fottrell (eds): Perspectives in Coeliac Disease. Baltimore: M.T.P. Press, pp. 169–182.

Strober, W (1981) The regulation of gastrointestinal immune responses. Immunology Today *2*:156–161.

Zilko, PJ, Dawkins, RL, Holmes, K, and Witt, C (1979) Genetic control of suppressor lymphocyte function in myasthenia gravis: Relationship of impaired suppressor function to HLA-B8/DRw3 and cold reactive lymphocytotoxic antibodies. Clin. Immunol. Immunopathol. *14*:222–230.

Author Index

Subject Index